Elementary Particles

Fundamental Theories of Physics

An International Book Series on The Fundamental Theories of Physics: Their Clarification, Development and Application

Elementary Particles

Mathematics, Physics and Philosophy

by

I. Yu. Kobzarev
Institute of Theoretical and Experimental Physics,
Moscow, U.S.S.R.

and

Yu. I. Manin
Steklov Mathematical Institute,
Moscow, U.S.S.R.

KLUWER ACADEMIC PUBLISHERS
DORDRECHT / BOSTON / LONDON

Library of Congress Cataloging in Publication Data

Kobzarev, I. I͡U. (Igorʹ I͡Urévich)
Elementary particles / by I. Yu. Kobzarev and Yu. I. Manin.
p. cm. -- (Fundamental theories of physics)
Includes index.
ISBN 0-7923-0098-X
1. Particles (Nuclear physics)--History. 2. Particles (Nuclear physics) I. Manin, I͡U. I. II. Title. III. Series.
QC793.16.K63 1989
539.7'21--dc19 88-37306

ISBN 0-7923-0098-X

Published by Kluwer Academic Publishers,
P.O. Box 17, 3300 AA Dordrecht, The Netherlands.

Kluwer Academic Publishers incorporates
the publishing programmes of
D. Reidel, Martinus Nijhoff, Dr W. Junk and MTP Press.

Sold and distributed in the U.S.A. and Canada
by Kluwer Academic Publishers,
101 Philip Drive, Norwell, MA 02061, U.S.A.

In all other countries, sold and distributed
by Kluwer Academic Publishers Group,
P.O. Box 322, 3300 AH Dordrecht, The Netherlands.

Printed in The Netherlands

Contents

Preface

This book has come into being as a result of scientific debates. And these debates have determined its structure.

The first chapter is in the form of Socratic dialogues between a mathematician (MATH.), two physicists (PHYS. and EXP.) and a philosopher (PHIL.). However, although one of the authors is a theoretical physicist and the other a mathematician, the reader must not think that their opinions have been divided among the participants of the dialogues. We have tried to convey the inner tension of the topic under discussion and its openness. The attitudes of the participants reflect more the possible evaluations of the situation rather than the actual views of the authors.

What is more, the subject "elementary particles" as dealt with in the dialogue stretches over (2-3) 10^3 years of historical time and a space of $10^{6\pm1}$ pages of scientific literature. For this reason, a complete survey of it is unachievable. But, of course, every researcher constructs his own history of his science and sees a certain list of its main points. We have attempted to float several possible pictures of this kind.

Therefore the fact that Math and Phys talk about the history of elementary particles is not an attempt to present the scientific history of this realm of physics.

Our text does not claim to give an entire list of sources nor to unravel the complicated skein of contributions of varying scale and importance which confronts the historian in his attempt to explain how scientific results are in fact obtained.

The views of Phys are an averaging of the ideas of the professional theorist on how the theory of elementary particles came to its present-day state. Phys refers to some papers because he has at some time read them and they appear important to him, because everybody quotes them, or because they are referred to in books from which he was taught. The psychological standpoint of Math is somewhat akin to platonism, a viewpoint that mathematicians and many theoretical physicists subscribe to, including Einstein and Heisenberg in later life. The clear and detailed vision of "formal reality" that Math's professional training has earned him imparts a normative tinge to his utterances;

he would like elementary particles to be such and such, and therefore they will have a nicely formed theory. Exp and Phil in the main provide the necessary rejoinders.

In the comments to each dialogue, detailed references and quotations are provided.

The second half of the book "The Structure of the Theory of Elementary Particles" is supposed to have been written by Math. This is the fruit of the assiduous work of a man who has endeavoured to collect and write down the most important and firmly established facts and ideas as seen at the present moment. This is something that we can firmly lean on. Unfortunately, we can't go anywhere on this basis. The spirit of scientific and theoretical physical research lives in the preprints that fill the tables of the reading room of the library attended by Math; there is no kind of mathematical order in these preprints and everything that has been firmly established is of no interest to the authors of them.

This book was more or less completed in 1982 when the discovery of the $W^{\pm}$ and Z^0 mesons brilliantly confirmed the applicability of the quantum field paradigm up to energies of order 100 GeV and, correspondingly, up to distances of 10^{-16} cm. During the five years that have elapsed since then, there have not appeared any empirical bases for doubting our paradigm , or as they say at the conferences,"the standard model is in brilliant shape". Have there been any sort of shifts in our understanding of it ? One of them is beyond doubt. The development of quantum field theory on lattices has brought us face to face with the visual reality of quantum fields fluctuating in space-time. Thus the long-standing necessity to represent quantum fields as an assembly of weakly coupled oscillators, which has lain at the basis of perturbation theory, has fallen away: we have learned to "see" quantum fields outside these narrow limits.

For the theorist working in elementary particle physics the first prize has always been the correct guess as to how the next layer of reality not yet investigated by the experimentalists will look. A number of future projects, such as technicolour, a conjecture about the existence of the next level of spectroscopy, that is, "smaller" and more strongly bonded particles, called preons or some such, are entirely within the framework of existing ideas both in the paradigmatic and even in the technical scheme of things. One can find out about the ideas of this scheme, for example, in the article O.W. GREENBERG, "A new level of structure", *Phys. Today*, Sept. 1985, Vol.39,No.9,22-30.

The most decisive attempts to go beyond the bounds of the paradigm are the work on string theory in 10- or 26-dimensional space. This approach goes back to the work of J. Scherk and John Schwarz in 1974 and leads to the hypothesis that the observable world of 3+1 dimensions arises as a result of the compactification of the extra string dimensions up to Planck length of 10^{-33} cm. The present state of this theory can be ascertained from the book: M.B. GREEN, J.H. SCHWARZ and E. WITTEN, "Superstring Theory" in 2 vols., Cambridge Univ. Press, 1987.

The emergence beyond the limits of QFT consists primarily in the idea that the primitive object of the theory is not the point particle, but rather the extended one-dimensional object, the string. Meanwhile, the connection between this amazingly interesting and many-sided branch of mathematical physics with experiment has not yet been made. It is suggested that the supergravity, referred to in the Dialogues, arises as an approximation in the Schwarz-Green string theory. At the same time this would ensure the absence of the divergences that so worried our Phys. To give the reader a more detailed idea about quantum strings, we have included in this book as an appendix, an article by one of the authors called "Strings", written for the journal *Math. Intelligencer* in 1987.

Whether this whole complex of ideas is crazy enough to lead to real progress in real physics remains at present, in 1987, unknown.

However, many theorists believe that we are already ready for the breakthrough to a new level of understanding of the world, comparable in significance with the creation of quantum theory and the theory of relativity in the first quarter of the 20th century.

June 1982 - December 1987

CHAPTER 1

Dialogues

PARTICIPANTS: MATH - the Mathematician, PHYS - the Theoretical Physicist, EXP - the Experimental Physicist, PHIL - the Philosopher

TIME: Summer-Autumn 1982

Dialogue 1

PHIL. In recent years I have read a good deal in various popular accounts about how great changes have been occurring in physics. Several new theories have emerged, such as the Glashow-Weinberg-Salam theory of the electro-weak interaction, and the theory of strong interaction. Recently I read an article by Salam [1]; as I understand it, he reckons that physicists are now close to realising the "Einstein dream"; I refer to the unified geometric theory describing "all of nature". (To Math) Recently I saw your copy of the *Lucas Lecture* by Hawking [2] in which he expresses the hope that the complete theory has already been discovered and that it merely remains to solve it; so the end is quite close. Is this the case, and what sort of a theory is it ?

PHYS. Generally speaking, much of what you have mentioned is actually true. Indeed, theories of the electro-weak and strong interactions have been constructed and the first of these is undoubtedly correct; at any rate, all physicists think so [3]. It seems to me that the theory of the strong interaction is also correct, namely quantum chromodynamics or QCD. But to begin with, it is probably worth while having a talk about the phenomenology; in other words, what range of phenomena do these theories concern themselves with? Incidentally, there is the historical side as well: after all, belief that a final theory is close at hand is not new.

MATH. As I understand it, at a certain level there are a fairly small number of "fundamental particles" to describe the functioning of Nature, including the sun, surrounding us, so to speak, "closest to us". It suffices to take the *u*- and

d-quarks, the e-, ν-leptons, the $W^{\pm}$- and Z^0-bosons and the γ-quanta and their interactions. The next two generations of quarks and leptons will have to be made with a lot of hard work in accelerators. They are, perhaps, important for the early stages of the evolution of the universe, but not now.

PHIL. Yes, I read somewhere that one of the two third generation quarks has yet to be discovered.

MATH. Yes. As always, the theoreticians are saying that it is too heavy. But for this, we could see our way clear somehow. One way or the other, I imagine that if we can manage with the shorter list u, d, e, ν, γ, then in all likelihood we could have managed with the list P, N, e, ν, γ, but then somewhere in the nineteen thirties, it should also have seemed that we knew everything.

PHIL. That is a remarkable conjecture, but what does Phys say ?

PHYS. I should say that if we are to think in a consistent way, then we still need to know where the interactions are to be taken from. The W^0- and Z^0-bosons that you referred to are necessary in the modern theory; they are the carriers of the weak interaction. Gluons are needed to carry the strong interaction. However, the quantum field theory of the last thirty years has not yet percolated into the consciousness of all physicists, so perhaps you are right.

MATH. I have checked my conjecture myself; it has virtually been confirmed.

PHYS. What did you do to check up on it ?

MATH. I set up an experiment. I went into the "Phys. Fac" library, found the "Elementary Particles" section and started to look. I found the excellent book "Electrons (+ and -), protons, photons, neutrons, mesotrons and cosmic rays" by Millikan [4]. The author was very insistent that the vast project of Democritus had at last been accomplished and that the "true" atoms have been discovered. Apparently Millikan regarded it as obvious that these are in fact the particles that I have listed. For some reason nothing is said about the neutrino, although the book came out in 1935 when Fermi's work on β-decay was already in existence.

PHYS. Which edition did you use ?

MATH. I got hold of the 1935 edition; apparently there are more recent editions, but I was interested in the state of the art in 1935 prior to Yukawa.

PHYS. Most likely Millikan did not consider the existence of the neutrino as proved. In 1935 he was no longer a young man and an experimentalist by profession. To study the theory of beta-decay and then to ponder about whether the theory was convincing, was hardly a natural business for him; and at that time there were no direct proofs of the existence of the neutrino.

MATH. I think it would be very interesting to consider carefully what is a "direct proof" if only for the case of the *P*, *N*, *e* and ν particles.

PHYS. What could be simpler ? The experimenter usually considers that he has obtained "direct proof" if it has been shown to him that as a particle "flies off" or "flies through", something has been "made", a neutrino, or a gluon. "Gluon jets" appeared and one began to talk about the "discovery" of the gluon.

MATH. Wonderfully simple ! But why, for example, did it take such a long time for the electron to be discovered and who actually discovered it ? I just cannot make this out.

PHYS. Well, of course, we must make some reservations here. Whether an electron or a gluon, the development does not always start off from the fact that the physicists "see" a new particle even though such things do happen. In quite a few cases one can make conjectures about the objects from circumstantial evidence; as in the Conan Doyle detective story, much is known about the criminal even before one meets him face to face. Last century, physics developed slowly and the electron was talked about long before it was given a name and the name was given before anyone had seen it.

PHIL. How can one talk about an object that does not have a name?

MATH. You know full well, of course, - by means of a descriptive phrase. Here is an example from Millikan - this is the German physicist, W. Weber, 1871 [5].

PHIL. I always thought that the planetary atom was invented by Rutherford.

PHYS. Was there anyone who *didn't* discuss the planetary atom prior to Rutherford ? The novelty of Rutherford's work [6] was that it presented experimental proofs of the fact that the positive charge of the atom is concentrated in a very small region.

PHIL. And where did the proof lie ?

PHYS. In the agreement of the formula, deduced by Rutherford for the scattering of α-particles by a nucleus with results of measurements.

PHIL. Then, surely, Rutherford's article should have opened with the words "I have managed to obtain direct proofs of the conjecture of the planetary atom considered previously, for example, in the works of ..." .

PHYS. Perhaps it did not begin like that. In any case, the then ethics concerning references possibly required nothing of this sort.

MATH. What do you mean ?

PHYS. Roughly the following: a reference by Y to the work of X, as it seems to me from my acquaintance with the articles of those times, means "I made use of the works of X", or even "I have developed the work of X". This sort of homage was not, on the whole, very readily given. Conversely, the absence of a reference meant "my work, as regards its method or its means of proof" - in other words " as regards its basic contents" is independent of the work of X, ...".

MATH. If you are right then this must make the work of the historians of science considerably more difficult if they wish to know which works and, more generally, which influences paved the way for, say, Rutherford.

PHYS. Yes, but Rutherford did not write his articles to make life less painful for the historians. But let us get back to the electron. (To Math) I recall very well the early history of the electron as recounted by Millikan.

MATH. And what was it like ?

PHYS. First of all there were the electric fluids of several kinds; for example, Franklin considered that electricity of one sign was associated with "ponderable" matter, and the other sign with "non-ponderable" matter. An alternative was the theory of two fluids *e* and *e*', the origin of which is described in Millikan's book by the somewhat obscure sentence: "Other physicists of the day, led by Symmer, 1759, preferred to assume that matter in a neutral state shows no electrical properties because it contains as constituents equal amounts of two weightless fluids which they called positive and negative electricity, respectively "[7].

As I understand it, the subsequent fate of the electric fluids turned out to be tied up with the fate of the atomic theory. In the first half of the 19th century, the atomic theory made enormous progress. Work was started by people who in fact were practically all physicists and chemists at the same time; the professions of the chemist and the physicist of that time clearly drew close to each other [8]. Read, say, Faraday's "History of the Candle": in it he emerges in the role of a pure chemist: he gives an account of simple and complex substances with regard to such things as oxidation and so on.

PHIL. Maybe the great physicist wanted simply to chat about chemistry. After all, "The History of the Candle" was a popular lecture.

PHYS. No. There were works by Faraday that nowadays would be classified as pure chemistry [9]. Gay-Lussac also had such interests; and Lavoisier, together with Laplace, measured the heat capacity of materials [10]. It seems to me that we should realise that the system of the sciences was being reorganised all the time: at the beginning of the 19th century, there existed rather the undifferentiated "science of matter" containing elements of both physics and chemistry. Subsequently chemistry separated off. A person such as Liebig was a chemist in the pure sense. I think the main point is that chemistry was the basic science of "matter" in the 19th century and up to the end of the 19th century it remained so to a certain degree. Certainly it is no accident that M. Curie and Rutherford obtained their prizes in chemistry; possibly it was not merely that in their works there was in fact a lot of chemistry [11]. Certainly the chemists generally regarded atoms as being their concern. If one can talk about a central problem of "the theory of matter" of the 19th century then surely one must regard this to be the problem of atomic weights. At the first international congress of chemists in Karlsruhe in 1860 there was, at last, agreement with regard to the atomic weights of the elements, although these were, of course, the relative atomic weights [12]. At that time there arose as a result of progress of the kinetic theory of gases the first values of the atomic numbers in the gram-molecule. In this context the planetary model of Weber (1871) and the article of Stoney (1874) are quite natural.

PHIL. And what did Stoney do ?

MATH. Stoney took the electric charge which must be passed through an electrolytic solution in order to obtain one gram-atom of a univalent substance. Then he divided this charge by the number of atoms in a gram-atom which was already approximately known at that time [13].

PHIL. What value did he obtain for *e* ?

MATH. 0.3×10^{-10} CGSE, roughly an order of magnitude less than the modern value of the charge 4.8×10^{-10}; the Avogadro number was not very well known.

PHIL. I once saw his article; I don't think he talked about particles.

PHYS. This may not have been by chance. In his1873 "Tractatus", Maxwell explicitly expressed the hope that in a true theory of electricity, molecular charges would disappear [14].

PHIL. That may be, but didn't the electron particle need to be discovered, that is to say, observed as a particle ?

PHYS. Of course.

MATH. Who, in fact, did ? I could not get it from Millikan, so I went back to von Laue's book [15] where I read that cathode rays were discovered by Plücker and that it was subsequently established via the researches of several people that they consisted of charged particles. "Under the influence of the brilliant experiments of William Crooks carried out in 1879, the hypothesis that cathode rays consisted of particles was firmly established, although Heinrich Hertz in 1883 wanted, on the basis of experiments that were inaccurate as a result of the inadequate experimental techniques, to regard them as longitudinal waves ...".

PHIL. What are longitudinal waves ?

PHYS. It was "known" from Fresnel's time that light consists of transverse waves in the ether; people were astonished that there were no longitudinal waves. But Hertz thought that cathode rays were longitudinal waves.

PHIL. And how was this possible if many people had established that they were charged ?

PHYS. Well, take von Laue. It says there that the early experiments were not convincing, and the question was solved by J. Perrin in 1895 and J.J. Thomson in 1897.

PHIL. And then what ?

MATH. Well then, according to von Laue, "from 1897, several researchers, among them Wien and J.J. Thomson, also George Fitzgerald (1851-1901) and Emil Wiechert (1861-1928) showed that in cathode rays the ratio of the mass to the charge of the particles was roughly 2000 times smaller than the hydrogen atom. Thus in decisive contrast to Hertz's ideas, they concluded that particles forming currents of anode rays are ordinary electrically charged atoms or molecules: conversely particles of cathode rays are "atoms" of negative electric charge, in other words, electrons".

PHIL. Were all these researchers you have mentioned independent [16] ?

PHYS. Wiechert published his results before Thomson. Thomson, it would seem, was independent of Wiechert.

PHIL. I once read a popular book where it was quite firmly asserted that the electron was discovered by J.J. Thomson [17]. What's the reason for that ?

PHYS. Very likely it is to do with the tendency for authors of popular books and text books to simplify the history. Many historians of science think that the notion of *discovery* should not in general be used by historians of science [18].

MATH. Another line of approach can be seen in von Laue. Somewhat strangely, and getting the chronology wrong, he wrote "At the end of 1896, when the Lorentz theory of the Zeeman effect (relating to the spectral lines caused by electrons in atoms) led to the same value for the ratio of the charge to the mass, the existence of the electrons was, after the efforts of 40 years, firmly established ....". What is this all about ?

PHYS. In fact, there was a further line of thought in the history of the electron connected with Lorentz, Larmor, certainly, and others [19]. Lorentz, starting from the 1870's, took up the theory of electromagnetic phenomena in matter and again introduced point charges (which Maxwell wanted to do away with) and electromagnetic fields in a vacuum acting on these particles. He himself regarded this as a partial reversion to the point of view of W. Weber. He considered the charges, which he called "ions", to be bound in atoms by an elastic force (oscillators). This enabled him to develop dispersion theory. When Zeeman discovered the splitting of spectral lines in a magnetic field [20], Lorentz gave an explanation of this and at the same time e/m was determined. In fact the Zeeman effect is more complicated than that obtained by the Lorentz theory, but the order of magnitude of e/m was correctly obtained. Thomson already knew this result in 1897.

PHIL. It seems to me that you and Math are somehow complicating the issue: after all, the Nobel committee awarded Thomson the prize for the discovery of the electron all the same.

PHYS. Not so; it was for the investigation of discharge in gases.

MATH. In the end this is not so terribly important. Apparently it would be correct to say that at the end of the nineties, it was established that charged particles in atoms were responsible for the radiation of light and that the particles of cathode rays were one and the same and that they were lighter: e/m of the order of 1/2000 of e/m for ionized hydrogen atoms.

PHYS. In fact e/m could be larger if the "Zeeman particles" and cathode ray particles had a very small charge, and J.J. Thomson gave consideration to this. However, in those years, of course, the Avogadro number was already well known so that, according to Stoney, it was possible to find the elementary charge and be satisfied that under any natural hypothesis, this was indeed the charge of the particles in atoms and cathode rays in which m is small. And that is how it was dealt with. At the beginning of the century a consensus was brought about: in atoms there are electrons. In 1904 H. Poincaré in his celebrated lecture said: "We know that the spectral lines of radiation arise from the motion of electrons. This is proved by the Zeeman phenomenon: that whatever is vibrating in a radiating body feels the influence of a magnet and hence has a charge" [21].

MATH. That is essentially a strange point of view. It follows from all that we have been saying that the important point is not that "whatever vibrates" has a charge, but rather that the ratio e/m is the same in cathode rays and atoms. And anything that is a particle radiating electromagnetic waves has a charge; that sort of thing is clear if the electromagnetic theory of light is true at all.

PHYS. You are certainly right, but when a consensus has already been established, the true sources of its origin are generally badly understood. By 1904, electrons had already been identified in the most varied sites: in β-rays of radioactive materials, in the photoelectric effect.

MATH. In that case, what was the value of the experiments of Millikan himself whose book we have been discussing ?

PHYS. In the direct measurement of the size of the charge of the electron. One gets the impression from the book by Holton [22], who goes into the laboratory write-ups of Millikan, that in essence, Millikan was at that time so confident of the answer, that he simply rejected those measurements in which the value of e obtained was not as desired.

MATH. What does "as desired" mean ?

PHYS. The Avogadro number was well known [23].

MATH. Then what was the "reality "?

PHYS. Measurements should not be done in that way. Millikan was lucky; if the "answer" which he knew beforehand had been incorrect, then incorrect work would have been performed. This sort of thing happens.

MATH. So Ehrenhaft simply displayed unwarranted trust in the facts ?

PHYS. Yes, something like that. It seems that the Millikan-Ehrenhaft method at that time did not make the measurement of e possible.

MATH. Shouldn't we say that the picture of reality that we see depends on our preconceptions ? Millikan believes in the electron and he sees it. Ehrenhaft does not believe in the electron, but believes in Mach and does not see it.

PHYS. The work that an experimentalist publishes is not yet conclusive evidence. And of course, any of his or her "ideological predispositions" rely on the confidence of the experimentalist in his or her results. The truth is ascertained, as in a court of law, by comparing the evidence of various witnesses.

MATH. But is the decision of the court the final truth ?

PHYS. The knowledge that the physicist possesses, namely, the knowledge of numbers and equations, is in the end verified by the technology. A new apparatus is devised and it works in the way it has been calculated to work: for instance, the knowledge of the nuclear constants enables one to calculate the position of the cores in which the reactor will start to work, and this is how it happens. The problems of "true knowledge" become very complex when discussed over the tea table, but when it is a question of real systems that work (or do not work), then everything becomes, if not simpler, then at least more concrete. If the reactor does not work, one needs to know why. This has actually happened and the causes were found.

PHIL. But can it happen that because of personal prejudices of somebody like Millikan the reactor will not work ?

PHYS. There are various cases like this. When work was begun on the nuclear bomb and reactors in Germany, the celebrated physicist Bothe measured the cross-section of the absorption of neutrons by carbon; he obtained an incorrect value for the absorption; it was too large. As a result, the Germans thought that carbon could not be adopted as the retarding agent, but that it was necessary to use heavy water D_2O. This was manufactured in Norway and the English saboteurs substantially slowed down its supply for the nuclear project which the Germans called the "Virus house". As a result, they did not manage to get a reactor going even by the end of the war. If it had not been for Bothe, then possibly the Germans might have come across plutonium earlier and the history of the second world war might have been altered.

PHIL. How does plutonium come into this, and how did it come to be known ?

PHYS. Plutonium is produced in reactors and can, along with uranium 235, serve as material for the bomb. The history of the German project is recounted in the book"Virus House" by D.Irwing (William Kimber, London, 1967); Bothe's error is described in the chapter "Fatal Error".

MATH. By the way, the story of the measurement of the charge of the electron e also did not end in the way that we have said; in Millikan's book he describes his later measurements. He improved his equipment and obtained $e = 4.770$ esu, whereas now e is equal to 4.803242 ± 14 esu.

PHIL. What is esu and what does $\pm$ 14 mean ?

PHYS. $\pm$ 14 is the standard deviation of the Gaussian distribution; and as for esu, this is a system of units that has now gone into disuse.

PHIL. And what happened with Millikan ?

MATH. In his measurements for determining the drops, Millikan used Stokes's law, according to which the force of the viscous drag acting on a drop is equal to $6 \pi a \eta$, where a is the radius of the drop and η is the viscosity.

PHYS. In fact, Millikan made a correction to Stokes's law which is valid for small a.

MATH. How then did he obtain 4.77 instead of 4.80 ?

PHYS. Well, I read or heard somewhere that Millikan handed over the measurement of η to one of his colleagues, who measured η incorrectly. All the same, Millikan himself was a superb experimentalist and in the end his data were processed correctly by making the proper adjustment to η and then obtaining the correct value of e.

MATH. And how was the error discovered ?

PHYS. It seems that somebody measured the Avogadro number N by a new method, by measuring the crystallographic structure by an X-ray structure analysis. The wave lengths of X-rays were measured in the diffraction lattices and a new value for N was obtained, given as $e = 4.80$ esu. It is most amusing that after Millikan and up to the crystallographic revolution, several people measured e from the minute noise in a current connected with the fluctuations of the current as a result of the discreteness of e, but still somehow came up with the value $e = 4.77$.

PHIL. And how did they manage to do this ?

PHYS. When an experimenter begins work he obtains systematic errors, Heaven knows how; he looks for them until there is agreement with the preceding authority and after this he stops looking for systematic errors and starts collecting statistics. Such errors are encountered again and again.

PHIL. Then can we trust your numbers that you talk about so much ?

PHYS. Numbers in articles, with great circumspection. But numbers in tables of the "Particle Properties" type are apparently fairly reliable [24]. How stable the data on the fundamental constants are can be seen in the book "The fundamental constants and quantum electrodynamics", by B.N. Taylor, W.H. Parker and D.N. Landeburg, Academic Press, New York-London, 1969. It is clear, for example, that according to the graph for $1/\alpha$, this number has gone beyond the bounds of error.

MATH. It would seem that physics, more or less like philology, is an open issue, to quote Umberto Eco.

PHYS. I think so; not quite to the same extent as philology but to some extent, yes. Science is created by human beings and human beings are a long way short of absolute perfection.

Dialogue 2

MATH. Today let's begin with the elementary particle concept. Nowadays it seems to be most natural for the theorist to define it as a quantum of some specified field, but this is the definition in the framework of the so-called "formal reality" of Vernadskii [1]: a collection of ideas on reality accepted by the science of a specified period. But, as you know, that is not all; ideas change, but electrons and photons "always exist" in some sense; they are invariant with respect to the changes that occur.

PHYS. It seems to me that Vernadskii's "formal reality" is almost the same as the "paradigm" of Kuhn (Phys has the satisfaction of being the person who found the right word). The paradigm of modern fundamental physics is quantum field theory (QFT) which we shall be discussing in detail later on.

PHIL. You have used the word "paradigm" in a somewhat unusual sense. As I see it, this is a linguistic term. Do you have a dictionary ?

MATH. (*Gets the book from the shelf*)

PHIL. (*Reads the definition of "paradigm"*) [2]

MATH. Since Kuhn, historians and physicists have used this word in a different sense.

PHYS. Apparently the time was ripe for introducing such a notion, a name was required and the word "paradigm" is now the accepted word, or so it would seem.

MATH. Names are not the point here; let it be "paradigm". But what happens to the objects when the paradigm changes ? Don't forget, none of us is insisting that the paradigm of QFT is the final one; it will change, but what will finally become of the electrons, photons, and quarks ?

PHYS. I don't see any problem here. The electron, of course, goes over from paradigm to paradigm; but this simply means that we recognize new aspects of its behaviour. Imagine a student X who sees Professor Y at lectures, examinations and seminars; he gets to know his lecturing style, takes exams, discusses scientific problems and even solves them to some extent. This is Y in the "professor" paradigm, with the symbolic extensions "strict", "clear" (or unclear) lecturing style, ingenious solver of complicated problems (or conversely, unoriginal thinker), and so on. He then meets Y on the tennis court and learns much more about Y as a tennis player. The paradigm now changes; X gets to know many new properties of Y. So it is with the electron. From the point of view of the theoretician the electron first lived a fully fledged paradigmatic life within the framework of the electrodynamics of Lorentz in which it acquired a charge, a mass, was a source of the fields E and H and was subject to the action of the "Lorentz force". This, so to speak, is the electron in the paradigm of Landau and Lifschitz's book [3]. It is already capable of doing many things, such as moving in an external field and radiating. Going over to the third volume (Quantum Mechanics) it has acquired certain new capabilities and forgotten others. (Non-relativistic mechanics is set forth in Volume 3.) Now, it cannot move with velocities close to the speed of light. Instead, it now possesses spin, does not always radiate, but is capable of diffraction.

In Volume 4, where quantum electrodynamics (QED) is set forth, it is once more capable of doing everything it could do in Volume 2, plus several new things, in particular, annihilation with positrons.

MATH. You say that the electron of Volume 3 "is not capable of moving with velocity of the same order of magnitude as the speed of light", but this is in fact not true. It can, but not correctly. You obtain the incorrect value of the energy $E = mv^2/2$, while in fact it is $E = mc^2(1 - v^2/c^2)^{-1/2}$.

PHYS. Of course, although this, strictly speaking, is not related to quantum mechanics. But we do know that non-relativistic dynamics cannot be used as $v \to c$. I don't see any problem here. All physical theories are approximate. In any case, we could have gone straight from Volume 2 to Volume 4; after all, the non-relativistic quantum theory is contained in the relativistic version.

MATH. But all the same, the theoretician, not knowing about the relativistic restrictions, would cheerfully predict that the electron could move with velocity $2c$ and have energy $2mc^2$, and this would be incorrect.

PHYS. We do not know in advance the limits of applicability of our theories; these are shown to us by the more extensive theories that come next, or by experiment. There is nothing particularly terrible about this, and this is how it always is. I should imagine, for instance, that electrons in biological systems in specific situations where one is dealing with phenomena related to consciousness, are not described by the theories set out in Landau and Lifschitz; if this were the case, then a theory of nature that included a description of the phenomena of consciousness would give new, at present unknown, restrictions on the applicability of QFT (it goes without saying that QFT includes all the preceding theories).

MATH. It seems to me that the majority of physicists take the view that the brain is to some extent like the computer and can be described even in the classical language, let alone the quantum one.

PHYS. Here there are differing opinions [4] and to be sure, we cannot come up with the right guess by counting the opinions. If we return to the electron, then it seems to me that nothing bad has happened to it in its transitions from paradigm to paradigm.

MATH. Provided you don't take into account the fact that it has lost its "substantiality"; it can now disappear under annihilation. The electron of QED is no longer a "thing", but a "quantum of a field"; it can come into being and disappear. In essence, we know nothing about it, apart from the formal rules of calculation of cross sections and instructions to the experimentalist.

PHYS. Yes, this apparently is a characteristic experience. We describe people whom we encounter in life by means of a paradigm constructed within the framework of natural language; but we manage to make the reality of this person whom we see at the lecture richer than the "professor" paradigm; he has a material existence that we perceive with our senses. But with regard to the electron, this is not at all the same; what we can say about it has to be said in the language of our "symbolic generalizations". There is the agonizing feeling that we know practically nothing about it. Sommerfeld complains about this in his famous book [5]: "with regard to the electron itself, there is little that we can properly say about it". Funnily enough, he wanted to "draw the electron" just the same, and he drew a point from which lines of force emerged. One ought to be more amazed at the fact that knowing so little about the "electron itself" we can say so much about bodies that are constructed from electrons and nuclei.

MATH. But the electron of QED, whatever else it may be, is not a "thing".

PHYS. Certainly it is not a thing in the sense that applies to chairs, tables or stones. A rock lasts longer than a human being. We can go back to the mountains and see a stone that we saw forty years ago. All this is connected with the surplus of baryons in the visible part of the universe: there is nothing around us with which we can annihilate the protons.

PHIL. Nor the electrons; but physicists for some reason are nowadays talking all the time about protons rather than about electrons.

MATH. This is because they now think that protons can decay, say, according to the scheme $P \rightarrow e^{+} + \pi^{0}$; the situation is then not so good; everything is disappearing although not very quickly.

PHIL. How so ?

MATH. The protons need to be re-made from time to time.

PHYS. Let's forget about innovations for the moment. After all, the annihilation $e^{+} + e^{-} \rightarrow 2\gamma$ itself already points to the fact that elementary particles do not possess the continuous existence that is enjoyed by ordinary objects; we must reconcile ourselves to this. Photons, on the whole, have a predisposition to be created and annihilated.

PHIL. Why is it that the electrons always turn out to be the same in this process of destruction and creation ?

PHYS. In a certain sense we don't understand this, but can only describe it. In any case this is embedded in the very foundations of QFT. The electron is always the same because the equations of quantum field theory do not change, while quantum mechanics certainly makes sure that the quanta always have the same charge and the same mass. This is defined by the constants in the equations.

PHIL. Can't the constants change ? For instance, can't they depend on time ?

MATH. Certain of the constants have now already changed. For instance, according to unified theory of the electro-weak interaction (QFD), the mass of the electron is now, in fact, the result of the interaction with some external field, and under certain conditions this external field may vanish, in which case the electron becomes massless [6].

PHIL. Where does the external field come from, in other words, what is its source ?

MATH. Ask Phys.

PHYS. According to hypothesis it does not need to be in a source, but exists in its own right; it arose primordially as the universe was cooling down.

PHIL. I have never heard about this type of field.

MATH. They have only recently been thought up. The theorists have altered their equations a lot.

PHYS. I think the answer is essentially still correct, all the same. Some of the constants have ceased to exist as a result, but then new ones have emerged.

MATH. I think we shall have more to say about the unified theories; it seems to me that even within the framework of QFT, all is not clear. If the fundamental objects are so "tethered to the paradigm", then to what extent do they "exist"? Besides, we have already come to the conclusion that these objects are not "things".

PHYS. I think that the one basic achievement of physics of the last century and this one has been that it has taught us that "conservation", so to speak, does not necessarily presuppose the existence of the things conserved. The ancient atomic theory took precisely the same point of view [7], as did the scientists of the 17th and 18th centuries. Once you have something that is conserved, look for "matter". Even heat was matter. The greatest revolution occurred when it was realized that what is conserved is not necessarily matter but integrals of the motion, say, energy. It would seem that gradually, everything that is conserved becomes an integral of the motion. When two electrons collide, the individuality of each of them is lost by virtue of their indistinguishability, but the charge is always equal to $2e$. When they separate, they reform the separate charges e ,e. In a microscopic piece of iron all the collective properties of the electrons are preserved, along with their total charge, but not the individualities of the separate electrons.

MATH. But if in QFT electrons and photons are merely terms that occur in certain sentences, while in the final event the question is one of predictions of experiments, then might it not be that under the next change of the paradigm these concepts and objects would disappear ? For example, the ether has disappeared ! To be sure, all kinds of things have disappeared, such as phlogiston, magnetic fluids, and so on.

PHYS. In the early stages of any theory there are, of course, hypothetical elements which can indeed disappear. But a developed theory is in some sense a direct description of facts and here, of course, nothing changes. Incidentally, this is well interpreted by Vernadskii who wrote that certain areas of "formal reality" cease to change when they have attained a certain level of truth [8].

MATH. And then one starts all over again. The general theory of relativity showed that Newtonian mechanics is invalid; there is no interaction at a distance, there are no forces. There is Riemannian space-time and geodesic lines of planets and other bodies.

PHYS. Such views have been expressed, but I think that they are naïve. In weak gravitational fields, one can to a first approximation, introduce the conserved momenta of the planets; in the next approximation, the momenta are variable; the change of momentum is determined by the position of the sun and the other planets; this is what forces are, and so on. Within the framework of the general theory of relativity you recover Newtonian mechanics to some degree of approximation. But the basic concepts have now been demoted in rank; they have become secondary constructs rather than primary ones.

MATH. All physical theories are phenomenological. Primitive concepts merely mean that at a given stage of development they are irreducible. Thus you must always expect that all the concepts we use will sooner or later lose their primitive status.

PHIL. Aren't you taking too gloomy a view of life ? Many physicists hope that a definitive theory will be created.

PHYS. On the whole, hopes of this type are never realized. Be that as it may, we are clearly concerned here with phenomenological theories, or conjectures.

MATH. But once a hypothetical element is assumed, where are the guarantees that a given theory has already achieved the respectable status of phenomenology and does not contain a conjecture ?

PHYS. We do not know but we have to act; in a sense, there is no rule here, but there is the individual possibility for the researcher to draw conclusions from all the facts available to him; and there is the collective consensus of scientific society. And nothing is absolutely reliable. Practically all physicists believed in the ether, and they were wrong. And don't forget that even the most cautious people spoke of "A conjecture that is all but proved" [9]. Newton made his conclusions about the validity of the gravitation of celestial mechanics by working with a relatively small number of phenomena — and he was not mistaken. Nowadays, when cosmic sputniks move along computed trajectories and successfully home in on Mars and Venus, we can hardly doubt the fact that the field of attraction in Newtonian space-time is, in a definite approximation, as accurate a picture of reality as that of a continent on a map.

MATH. Of course all this is true but even here, arguments break out from time to time about, say, "the reality of inertial forces".

PHYS. These arguments are blatantly spurious because in natural language you can, as it were, "understand" sentences that are in fact meaningless in a given theory. Newtonian mechanics is literally valid in inertial systems, because in this case there are no inertial forces. In a non- inertial frame, one can preserve Newton's second law by adding "inertial forces". They "exist" in a non-inertial frame if we wish to preserve the equation for the acceleration. The correct formulation would go something like this: "if we want Newton's second equation to preserve its form in a non-inertial frame, then we must add to the right hand side of the equation a certain term called "inertial force" over and above the usual forces".

MATH. But you know that in the general theory of relativity, gravitational forces are, properly speaking, also inertial forces.

PHYS. In the general theory of relativity the semantics is different. There the inertial frame is simply that of free fall.

MATH. But if existence depends on the semantics adopted, then could it not be that electrons, protons and quarks would disappear ?

PHYS. I don't think so. But in any case, in a wide class of situations we can talk about electrons and protons in everyday language as we talk about ordinary objects. And I doubt that we will reject this. Properly speaking, if we start with the electron, physicists were speaking about charged particles long before the first theoretical paradigm happened.

MATH. And what about quarks ?

PHYS. In a certain sense we can almost "see" them. If we scatter an electron or a neutrino at a large angle with energies of the order of tens of GeV, we have a resolution, as in the microscope, that is better than the distance between quarks and a proton.

PHIL. So what is visible ?

PHYS. The wave scattered at the point centre .

PHIL. The wave of what ?

PHYS. Let's say, the de Broglie wave of the scattered electron.

PHIL. And why do you say that you can see the quark ?

PHYS. Well, the thing that does scatter the point centre is observed.

PHIL. But you see the scattered wave, not the quark.

PHYS. All right ! I don't see you either, but only the light scattered by you.

Dialogue 3

PHIL. In our first discussion we somehow got along splendidly without paradigms. How did we manage this ?

PHYS. We can begin by saying that the paradigm was present behind the scene. In fact, of course, J.J. Thomson and his contemporaries calculated the motion of an electron in electric and magnetic fields by using the formulae for the fields of magnets and capacitors, the Lorentz force and Newton's equations. Here the electron is featured as a typical "material point with mass and charge".

MATH. But don't forget that electrodynamics is invariant with respect to Lorentz transformations, while mechanics is invariant with respect to Galilean transformations; doesn't this mean that the paradigm was in fact contradictory?

PHYS. Formally there is no contradiction. It is simply that the frame of reference in which one has to work is singled out.

MATH. But, as you know, the earth moves "relative to the ether".

PHYS. Yes, but in essence, this speed is small relative to that of an electron in a cathode-ray tube, so that in fact, they didn't have to bother about this. In reality, they most likely didn't think about this, but knew that the electrodynamics of capacitors, magnets and charges works in a laboratory.

MATH. Lorentz thought about these problems.

PHYS. Yes, because he took the Michelson experiment seriously and attempted to explain it [1].

PHIL. Why did you say "We can begin by saying..." ?

PHYS. Because physical theories are not like mathematical theories, that is, a more or less ordered system of postulates, definitions and corollaries, as in Euclidean geometry. Rather, a theory is a fund of images, ideas, even associations used to explain a region of phenomena.

MATH. But you yourself say that the fundamental paradigm of modern elementary particle theory is quantum field theory. Well then, surely this isn't so chaotic.

PHYS. You know, of course, that modern quantum field theory is by no means a beautifully-proportioned palace, but rather an old house, all overgrown with extensions put together. Of course, the original central structure remains intact, the particles are obtained by quantization of fields, but there are so many outhouses !

MATH. That's true: from vacuum expectations to quantization along contours - a fairly formless structure.

PHYS. That is more or less how it has always been. The attempts to order the subject in a logical fashion, as it were to give an ideal pure structure to physics have rarely borne fruit; this is more the preoccupation of the natural philosophers. Democritus developed a remarkable paradigm for ideally solid atoms. After the introduction of central forces in mechanics, Boskovich [2] tried to settle the "paradigm of central forces", but the physicists were much more careless. With Abraham [3] (approx. 1900), the electron became, say, a small hard sphere.

MATH. So Abraham mixed Lucretius with Boskovich ?

PHYS. Roughly speaking, yes. This mixture apparently looked attractive.

PHIL. Why?

PHYS. Sommerfeld [4] thought that the rigid ideal sphere with surface charge was, as a model of the electron, a perfectly natural element of the "electromagnetic picture of the universe". In any case, the postulate was simple and enabled one to perform calculations.

MATH. But such an electron contradicts the principle of relativity.

PHYS. At that time the principle of relativity seemed to Sommerfeld to be a mechanical atavism. There were more complex models of the electron. In order to construct an electrodynamics completely consistent with the principle of relativity, Poincaré introduced an electron with an ideally stretchable charged surface, which was under a uniform constant pressure from the ether [5].

MATH. Doesn't this model reduce to the Lorentz-invariant action ?

PHYS. Certainly. If we have an electron at rest with constant surface density, then it will be a sphere, while if it is moving "relative to the ether" it will undergo the Lorentz contraction.

PHIL. What happened to all these models ?

PHYS. Planck wrote down the Lorentz-invariant action for a point electron, and there was no longer any need for the models [6].

PHIL. Don't forget, as I recall it, the electromagnetic mass is infinite for a point electron.

PHYS. It was implicitly agreed to consider only problems on motion in an external field and on radiation; not on the action of the field of the particle on itself.

MATH. It isn't quite like that. In classical electrodynamics there is a force of radiational friction [7].

PHYS. That's a very crude approximation; it is the first term of an expansion in which the following terms are meaningless.

MATH. Nevertheless, this was taken up in a big way in the 30's.

PHYS. Yes, but nothing satisfactory was obtained and the question no longer arose, since the method of renormalizations in QED in practice solved the problem of calculating the radiation corrections.

MATH. All right then; so you are back on your favourite paradigm: it seems to me that you are almost as fond of it as Boskovich was of his central forces.

PHYS. Of course not, but it is the best paradigm available.

MATH. What can be said about the history of the proton ?

PHYS. It started with Prout who at the beginning of the 19th century suspected that everything was built from the same light element, hydrogen [8].

MATH. Superbrain Sherlock Holmes does it again !

PHYS. Or at any rate a piece of good fortune. When atomic masses were being measured, they did not quite turn out to be multiples of the atomic mass of hydrogen, so this idea was dropped at that time.

PHIL. And what resurrected it ?

PHYS. The discovery of isotopes, Rutherford's planetary model and the explicit determination of the atomic masses of isotopes by the method of mass spectroscopy in 1900-1918. After this, as it turned out, the suggestion that the atomic nucleus consists of protons and electrons began to emerge.

MATH. Weren't mass defects a nuisance ?

PHYS. No ! When they thought about these things, they already knew the connection between energy and mass and the fact that the electromagnetic energy of a densely packed system of charges could alter the mass seemed quite natural [9].

MATH. Is there a "first" author or a list of first authors who suggested P,e as the model of the nucleus ? What do the text books say ?

PHYS. It seems that they did not get round to mentioning this. Possibly one of the first authors was a man of great originality, the Dutchman van den Broek [10].

MATH. Is it true that the existence of protons was recognized even before they had actually been observed ?

PHYS. No, because proton beams were easy to obtain and J.J. Thomson and others had clearly "seen" them. In fact, the photon and the neutrino have a more interesting history. These particles were invented by theorists. The photon was, so to speak, "invented" by Einstein; by "experimenting" with Wien's formula he discovered that at the Wien limit, black body radiation reduces precisely to a gas of particles: the fluctuations of the number of particles follow a $\sqrt{n}$ law; thereafter, he discovered a little later by means of thought experiments that the photon also has a momentum [11].

MATH. In fact, in this second work he already was using Planck's law and already was on the track of the "wave aspect".

PHYS. That is true, although he explained it incorrectly.

MATH. Well, of course, that's also true. But don't forget that one should say that the photon is not a particle proper but a quantum of a field, that is, an elementary excitation of a quantum system and that sort of thing, and since Einstein did not know about quantum mechanics, he had to put forward some erroneous construction.

PHYS. You are right, of course, but the basic difficulty was the impossibility of explaining interference and it made us all think that the photon conjecture was irrelevant [12].

MATH. So presumably, in order to accommodate light quanta, one had to wait for a quantum theory of the electromagnetic field?

PHYS. No ! As usual, the physicist did not wait until all the contradictions had been got rid of. I think it was Compton [13] who convinced everybody by showing that the momentum of hard quanta in this particle scattering by electrons, changes in accordance with the laws of elastic collision. After this, people believed in quanta and began to look for theories that would seriously accept the dual nature of light.

MATH. Fine. And what about the neutrino ? I have looked at Fermi's article, and I think that was tremendously lucky. It was, so to speak, directly given birth to in its paradigm. Fermi writes so thoroughly about the fields of the electron and the neutrino, second quantization and the creation-absorption operators, that the connection between the paradigm and the object is strikingly clear.

PHYS. Anyway, the object appeared in a more informal guise. The neutrino was thought up by Pauli in order to solve the difficulties connected with the non-conservation of energy and angular momentum; so that to begin with, this was simply a "particle" and Pauli thought it to be fairly massive and housed in the nucleus. In other words, his neutrino was at the same time almost like a neutron [14]. It needs to be said that things worked out better for the neutron. The neutron was comparatively easy to see. After Chadwick's work, where essentially it was shown that the neutron was a heavy particle with mass roughly the same as that of the proton and strongly interacting with nuclei and protons, it immediately received universal recognition [15]. As regards the neutrino, after it had been theoretically discovered by Pauli and Fermi, many years passed, until Reines in 1957 managed to observe the reactions caused by neutrinos (more precisely antineutrinos) generated in reactors, and after this it became clear to everybody that this was a particle just like any other [16]. Nowadays, of course, the observation of reactions caused by neutrinos is everyday practice. The absence of ν in Millikan's book [17] is most likely a reflection of his mistrust in the theorists.

MATH. The positron, I take it, enjoyed a happier fate ?

PHYS. Yes, this was discovered by experimentalists in the usual way, and only after this was it realized that it was an antiparticle, the existence of which follows from Dirac's quantum mechanics of the electron.

MATH. Wouldn't it be better to say from the quantum theory of the Dirac field with spin 1/2 ?

PHYS. Yes, but this formulation came later. Dirac proceeded more intuitively: he filled up the levels with negative energy. The modern symmetrized picture of e^- and $e^+$was a later achievement [18].

MATH. That's fine, but it seems that the picture in which discovered particles are straight away interpreted within the framework of an already established quantum field theory cannot be a correct one. Otherwise, how can one explain why Chamberlain and Segre were awarded the Nobel prize in 1959 for the discovery of the antiproton ? If the field theory were such an obvious thing, then you might as well say that once there is P, then we can talk about $\overline{P}$ as well.

PHYS. I think the picture of the development of science as a succession of paradigms, as suggested by Kuhni, is a fairly oversimplified one. In the thirties, it turned out that the difficulties in QED were very great; it was almost, as it were, not a theory. Therefore, there had to be many people ready to agree that in general there are neither antiprotons nor antineutrons. In general, what a given researcher believes in or does not believe in is a very individual matter. Anderson, when he discovered the positron, did not wish to deal with the Dirac equation. On discovering the positron he wrote that he had discovered a state of the proton with large radius [19].

MATH. From symmetry considerations, he should all the same have believed that there are heavy "electrons" with small radius.

PHYS. That is very plausible. Maybe there is something about this in his 1932 article. Somehow or other, even in 1958, there were certainly many people who had doubts about the applicability of Dirac's equation to the proton and consequently doubted the existence of $\overline{P}$.

MATH. What is extraordinary is the fact that, on the whole, they were right. The proton, as we now know, is not elementary, but is a bound state of three quarks, and the Dirac equation, as far as that goes, has really got nothing to do with it.

PHYS. It has nothing to do with it in the sense that its magnetic moment bears no relation to Dirac's $e/(2m_p)$, but all the same, we use the Dirac equation in the calculation of eP scattering, although we do introduce an anomalous magnetic moment and two form factors.

MATH. So what properly remains of the Dirac equation ?

PHYS. Very little in actual fact; only the property of the states of particles with total spin 1/2 under Lorentz transformations. In a certain sense, when we use the Dirac equation in scattering calculations for the description of the states of the proton in a space free of an electromagnetic field, we are merely paying homage to tradition. In essence, all the results could have been obtained from the properties of Lorentz invariance.

MATH. That's fine. But all the same, modern field theory with internal degrees of freedom that are not related to the Lorentz degrees of freedom, spontaneous breakdown of corresponding symmetries and elementary quarks and gluons are already quite different from the quantum electrodynamics of the 30's. How did the transformation take place?

PHYS. Little by little. It all began with the nuclear forces: it was discovered that the nuclear interactions PP and PN, if these are taken to be in the same state, are the same. Thus there came about the idea that P and N are two states of one and the same particle and the nuclear forces need to be written in the isotopically invariant form. The well-known spin matrices for particles with spin 1/2 were already being used, and by analogy, interactions with isospin matrices τ were introduced.

MATH. I think that you are oversimplifying the whole story to some extent. As I recall, the two states of the nucleon and the τ operators appeared earlier. They were already used in Fermi's article on the theory of β-decay in 1934 in which he makes reference to Heisenberg's work of 1932 [20]. I doubt whether in the year that the neutron was discovered nuclear forces were as well-known as all that.

PHYS. Oh yes. Even for Heisenberg, the fact that the masses of the proton and neutron were close was, it would seem, sufficiently strong evidence for him to regard P and N as "two internal quantum states of a heavy particle". The τ operators turned up because he wanted to consider the exchange forces for P and N when P is converted into N and N into P.

PHIL. Why did he want to do this ?

PHYS. Most likely he wanted to construct the most general form of *NP*-interactions. Somehow or other, after this, as empirical facts indicated that the *NP* and *PP* forces are the same in corresponding states, there followed the inescapable conclusion [21] that the energy of interaction must contain either the identity matrix or $\vec{\tau}_1\vec{\tau}_2$, where τ_1 and τ_2 acted on the isotopic variables of the nucleons 1 and 2 already introduced by Heisenberg.

MATH. I take it that everything here was described in terms of the usual group $O(3)$ rather than $SU(2)$.

PHYS. Yes of course. The situation was very confused after the discovery of new particles had suggested the idea of extending the group of isotopic transformations. The theoreticians experimented with the transition $O(3) \to O(4)$ but not with $SU(2) \to SU(3)$, which in fact worked.

MATH. Are you referring to the strange particles here ?

PHYS. Yes, but in the history of isotopic invariance, many interesting things occurred: Yukawa had suggested a field theory for nuclear forces based on exchange of mesons [22]. The heavy meson predicted by him was erroneously identified with the μ-meson discovered in the 30's. This encouraged belief in the validity of Yukawa's theory so that at the end of the 30's, meson theories of nuclear forces were taken up in a big way. The isotopically invariant meson-nucleon interaction was written down and the mesons of the nuclear forces became triplets of the isospin group [23]. In his lectures of 1944, Pauli reckoned that it followed from the properties of the deuteron that nuclear mesons were pseudo-scalar. So we see that his then favourite field theory is the modern phenomenology of π-meson interactions at large distances.

MATH. But surely the π-mesons are now non-elementary.

PHYS. If one of the nucleons is at a distance $l \gg \hbar/(m_\pi c)$ from the other, then we can ignore the dimensions of the π-meson and regard it as a point. The situation may be even better. In fact, both the nucleon and the π-meson have dimensions of order $\sim\frac{1}{2}m_\pi$, and therefore one can go straight in to using the theory in Pauli's book [24].

MATH. Has all this been calculated and are Pauli's arguments correct ? In fact, does the π-interaction correctly explain the quadrupole moment of the deuteron ?

PHYS. Probably nobody knows exactly. Somehow or other, the isotopically invariant interaction of π-mesons and nucleons came into the world. It was then suspected that some relations between the probabilities of the strong interactions could be obtained without resorting to the dynamics, but directly from the properties of the isospin group. And this is how we became acquainted with the first internal symmetry group. Starting with the first works of Leprince-Ringe and Butler and Rochester, which was done in the 40's, physicists began to observe new particles (*K*-mesons, hyperons) by studying Wilson cloud chambers and the photoemulsion levels created by cosmic rays [25]. These were called strange particles because they decayed into π-mesons and nucleons considerably more slowly than was expected.

PHIL. What were they expecting ?

PHYS. Roughly speaking, it was thought that particles that are properly created (and this was known), ought to participate in strong interactions, and since the final products were strongly interacting π's, and *P*, *N* 's, they expected the decay time to be 10^{-23} sec. They looked for explanations and thought about isospin again, but in the event, Gell-Mann in 1953 nearly gave the correct explanation; after this the isotopic properties of elementary particles started to be intensively studied [26].

MATH. And it has now become clear that, in essence, there is nothing fundamental about the group $SU(2)$. It is simply that the masses of the *u* and *d* quarks are certainly not equal to but much less than $\hbar/cr_c$, where r_c is the confinement radius . Therefore in the approximation m_u, $m_d < \hbar/cr_c$, they can be neglected. And the " isospin exclusions" are not fundamental either; it is simply that in strong interactions, quarks are not converted into one another; that is all there is to it.

PHYS. Certainly it has been very useful to have thought for 25 years that the isospin group was "very fundamental". Otherwise, Yang and Mills surely would not have started to try to make isospin symmetry local.

MATH. The situation in modern elementary particle theory is utterly weird. Those internal symmetry groups that are in fact "visible" and occur in observable data (isospin, and the group $SU(3)$ of the symmetry of sorts) have a random origin and are non-local. Nevertheless, it is thought that both the idea of internal symmetries and that of their localization is correct and there are local true symmetries $SU_c(3)$ and a group of weak interactions $SU(2) \otimes U(1)$, but they do not show up in the external world; that is because the first group is represented only by singlets while the second, in general, is broken spontaneously. The history of all this, as it turned out, was more appropriate for the comedy of errors than an orderly inductive process along the lines of Stuart Mill.

PHYS. Of course. But, in the end, at the basis of the conjectures that finally led to the modern theories there is simple inference by analogy. In QED there are phase transformations of the material fields (the group $U(1)$). These transformations can be made local, because the vector field also enters into the theory. If a symmetry group G is realized on material fields, then we can, by analogy with QED, construct the Lagrangian so that it will also be local [28]. Combined with the conjectures that led to the colour group and the weak group there are always elements of conjectured truth, erroneous identifications and preconceptions. In the end, the errors led to a contradiction with the facts and disappeared from the scene, while the fragments of truth were amalgamated into a coherent picture.

MATH. It seems to me that we hardly need to pore over the details of the history of the creation of the modern gauge theories at this point. To be sure "the day is not over yet". But I don't think the road from QED to the Yang-Mills field theories was as smooth as you have made it out to be. For doesn't it seem that at the end of the 30's, people were inclined to believe that there would be a rapid end to field theory, also in the 50's and in the 60's ?

PHYS. Certainly. The line of development that I have not actually traced out but, if you like, have attempted to indicate with a dotted line, has over many years frequently almost disappeared from view, while on the surface, sensational attempts have been made to play out quite different scenarios. The extension of paradigms to a new area is never a painless process; contradictions inevitably arise, and there are always crises; indeed the paradigm itself is always being reorganized and altered. When Maxwell and Boltzmann and their predecessors and successors applied mechanics to atomic phenomena, they constructed a kinetic theory of gases and thereby ran into glaring contradictions ! Here success could only be partial and, properly speaking, it was only quantum theory that made it possible to remove these contradictions.

MATH. Perhaps the situation is still the same even now ? Maybe we are poised for a new revolution.

PHYS. It gets harder and harder each time to carry out a revolution, since our store of knowledge which cannot be left out, is ever increasing. Einstein was fond of the saying: "Mankind never learns from experience, as the old mistakes are always being presented to him in a new light". But all the same, before discussing the present day situation, I would like to recall the earlier mishaps of quantum field theory.

Dialogue 4

MATH. So what actually has happened in field theory between 1929 and the present day ? Essentially, I cannot understand anything. On the one hand, you have convinced me that this is the paradigm of modern theoretical physics. In fact, the theories of the electro-weak and strong interaction which the modern generation of physicists consider to be a description of "an objective reality actually existing outside us", are undoubtedly the "quantum field theories", which can be said to be the sisters of the quantum electrodynamics of Heisenberg and Pauli created in 1929. On the other hand, if, say, we look at the texts that Pauli wrote between 1929 and 1946 inclusively, he has, as it were, turned very severe towards his children, as though they were his stepdaughters, and is forever awaiting their death; and not only he [1]. We can glance through the journals of the 30's and we find that the only theories discussed there are the ones that do not have a place in the paradigm [2].

PHYS. And were there many of them ? And what were they ? Also it always seems to me somehow that the probabilities of the various processes of type $e^+ + e^- \to 2\gamma$ were calculated just in the 30's. Haven't you somehow checked this by collecting statistics ? How many such articles have there been?

MATH. I haven't gone into this. I have only glanced at the journals when working.

PHIL. But tell me, are there other living creatures who perhaps can lead us to new works ! Could it not turn out that they all died long ago ?

MATH. Well yes and no. All the same, there is much concerning our paradigm that has not got into the textbooks, so that if you want to understand it in modern terms, you have to look in the old articles. But the new results are made in a different way. I think all this is because we have not yet arrived at the moment of dismantling our paradigm. When we start to rebuild an edifice, we have to give careful thought to the foundations, but as long as you are making a superstructure, you do not need to look for the foundations.

PHIL. All right then, extensions and outhouses _ that is always possible. But if your metaphor is to be taken seriously, then when the next superstructure turns up, the foundations may be unable to stand up and begin to creep.

PHYS. Certainly that actually happens. All through the 18th and 19th centuries, the theory of continuous media was constructed; then trying to follow this image, the physicists began to construct the wave theory of light, then thermodynamics, then Maxwell constructed his electrodynamics, and all the time everybody thought that they were working in the Newtonian paradigm, and that they were building phenomenological superstructures on top of the mechanics of central forces, that were correct independently of the actual law of force.

PHIL. How actually is Newtonian mechanics related to Fresnel, thermodynamics and Maxwell ?

PHYS. Thermodynamics was considered to be the phenomenology of the mechanical theory of heat: here we have a system of an enormous number of atoms in which central forces act. The law of conservation of energy follows from this [3]. Have a look, say, at Helmholtz's celebrated article. It was also supposed that the ether consisted of "atoms of ether", and from this there arose the optics of Fresnel and the electrodynamics of Maxwell as a phenomenological description of Newtonian ether. Poincaré still thought this in 1904 [4].

MATH. Don't forget we have already said that in general physicists are extremely unconcerned about their paradigms. Poincaré was a mathematician and was well accustomed to the idea that he should know what sort of an object he was studying: that is to say, to point out the initial postulates. But I think that in 1904 the physicists were in total disarray. Some, such as Mach, believed in a pure phenomenology, while others, such as Sommerfeld [5] believed in the electromagnetic picture of the universe.

PHYS. Certainly it wouldn't be right for me to take issue with you on this. But I would like to say that Mach was also essentially a philosopher. He was always striving to understand what the paradigm was in which physics was moving, and this led him to begin with to the extraordinary conjecture that the Newtonian paradigm was incorrect or incomplete; and then, when he began tentatively to grope for the truth, without setting up experiments, and without engaging in theoretical physics, he was led into a gross error, namely, the rejection of the atomic hypothesis.

MATH. Weren't there in Mach's standpoint certain purely ideological influences extraneous to physics ?

PHYS. The classical ancient atomism which we associate with Leukippus and Democritus but know about mainly from Lucretius, from whom it was studied in all likelihood by Newton and Gassendi, was not able of course to become a scientific theory, but turned into some sort of pseudo-scientific religion. Have a look at Lucretius. Everything essentially is directed towards the idea that by rational arguments, that is, appeal to facts and logic, one can prove that the world "consists of" indestructible atoms, and the aim was to explain that there were no immortalities, no need to dread the gods, death, and the like. All this, in essence, became the religion of the "Epicurean sect". When Christianity arose, providing an alternative "theory of the universe", there inevitably had to be a conflict between it and Lucretius and Epicureanism [6]. Christianity was not able to compete with them in terms of logic and clarity; the vehicles of this ancient rationalism needed quickly to be destroyed or ousted. The enemy was feared and Christian theology conceived a hatred of Democritus.

PHIL. Yes I am sure you are right. There is a remarkable poem by Blake. Something like "Mock on, mock on, Voltaire, Rousseau; your atoms are sand against the wind, which comes back and blinds you" [7].

PHYS. I see; and, of course, at the end of the 19th century religious people, who paid no heed to atomic theory, were able to rejoice as of old when they heard that atoms were the product of our imagination and possibly even matter as well. As is well known, Lenin [8] wrote on the relationship between the theological wing of philosophy and Mach and his followers. Even so, confessional attachments did not on the whole play a decisive role; in the main disbelief in the atomistic viewpoint was to do with the practical difficulties of the theory.

PHIL. And what were they ?

PHYS. Well, you must remember that the Newtonian paradigm, if it was to be taken seriously, contradicted everything in the world; there has never been anything more in contradiction with the facts than the equipartition theorem. Everybody wrote about this: Maxwell, whose researches were largely responsible for the creation of the kinetic theory of gases in 1860, the cautious Poincaré in 1900 and 1904, and even Gibbs, who, just at the time of the gravest doubts, had worked out the foundations of statistical physics [9].

MATH. And how did Gibbs and Maxwell overcome these doubts ?

PHYS. Somewhere there is a remarkable sentence due to Maxwell. The idea that kinetic theory may not be true is improbable because there are too many confirmations of it; but something important is missing.

MATH. If the theory contains contradictions, how can one work with it ? And all thinking that deals in reality rather than with its own outcomes, always moves between contradictions, but moves forward all the same. Since we have mentioned Mach let's recall one of his *bons mots*; the process of cog–nition is the process of adapting thoughts to facts [10]. This is indeed the case; it is almost a biological process.

PHIL. But for all that, how can one reason, if there is a contradiction in the original premises ?

PHYS. L.D. Landau would have said "Keep trying one way and another". Actually, the answer is simple; don't construct overlong deductive chains and avoid the "zone of contradictions".

PHIL. And how does one limit them ?

PHYS. Again, in some rational way or other. If you had seen in 1915-1925 that the old quantization was working well for stationary states, then you too would have suggested that classical mechanics, with the quantization recipe $\int pdq = 2\pi n\hbar$, works for stationary states, while for transitional states nothing is known and for the time being, we don't have to think about them.

MATH. But you then learn to quantize the hydrogen ion [11] and realize that this hypothesis does not go through.

PHYS. You then think up something else. And so you will go on trying until Heisenberg makes his desperate step: throwing out the coordinates of the electrons altogether, and writing the equations for the "tables", that is, the analogues of the Fourier components.

PHIL. Did that work ?

PHYS. Certainly. It worked remarkably well. In the end, Heisenberg suddenly caught sight of the fragments of the true structure.

MATH. Why not de Broglie ?

PHYS. If you like, you can say "and de Broglie independently". There was, of course, an alternative path, namely from de Broglie to the Schrödinger equation for a single particle, then for n particles. Only this line of development was not independently brought to a conclusion.

MATH. Is this opposition reasonable? Of course, I understand that the article by Born, Heisenberg and Jordan is virtually a complete quantum mechanics for a system of n-levels, but then there was no probabilistic interpretation in it and I take it that Born's work helped to find it.

PHYS. Least of all would I want to try to underestimate the value of wave mechanics or its role. But all the same, Heisenberg's work of 1925 hardly needs any changes today; after all it *is* modern quantum theory.

MATH. It seems that we have forgotten about quantum field theory.

PHYS. No. Actually we have talked about it in the language of metaphors. Its fate between 1929 and 1974 was somewhat similar to that of the atomic theory. After the initial successes there appeared difficulties, new areas of facts emerged that could not be described as originally within the framework of quantum theory, alternative programmes were put forward, both ultra-revolutionary ones and extremely phenomenological ones. The ultra-revolutionary programmes were not able to be converted into anything intelligible and the phenomenological programmes were not able to be developed sufficiently effectively. A small group of theorists patiently continued with quantum field theory. Gradually some new ideas were formulated and in 1967 there emerged the correct version of the electro-weak gauge theory [12]. Then eventually the modern version of the theory of the strong interaction was put forward.

PHIL. All this is reminiscent of the story by Anatole France. Do you recall: "They are born, they suffer, they die". Anyway, what were the difficulties ?

PHYS. Certainly it was the divergences. Almost as soon as quantum field theory emerged, it was discovered that the first non-vanishing order for the cross section of the processes was finite, but the following ones were infinite; the mass corrections were infinite and so on.

PHIL. What did they decide to do about this ?

MATH. If you take Heitler's well-known textbook [13] the impression is that they thought that QED was applicable only at very large distances, and at other distances it was thought to be somehow quite different. They were amazed, it would seem, that in certain processes the theory worked for too long. For example, the formula for the bremsstrahlung of electrons holds at electron energies $E \gg mc^2$. Why is this ? The answer is given by the Weizsäcker-Williams method. The process can lead to Compton scattering, and furthermore, the effective photons with energies less than mc^2 are essential.

PHYS. Many of the people who were writing articles at that time are still alive. Probably we should draw up a list of them and ask them all.

MATH. Yes, I am sure you are right. WIth regard to the counter-programme, there were various ideas: the quantization of space, going over to non-local field theories. And most likely some other ideas as well.

PHYS. Yes, it's not easy to remember now: what else was there? At that time renormalizations gradually turned up and in the forties there came the great breakthrough: in QED it became possible to calculate any order in quantum field theory.

PHIL. Sometimes it is said that the whole technique of renormalizations is a lot of conjuring tricks with infinites. How could people decide to accept that?

PHYS. Theorists who do something concrete are never as irrational as their colleagues say afterwards. If you look at Feynman's work at the end of the forties, he has modified the theory of small distances and has in fact calculated the finite and small corrections, and his renormalization procedure is in fact a treatment of the small corrections [14]. This is how it is, in general, in QED theory. In gauge theories with simple groups there are essentially no problems due to asymptotic freedom at small distances. There, as the distances tend to zero, the corrections are always small.

MATH. You are not quite correct. In the Higgs sector all the old difficulties are still there; the corrections increase as the cut-offs are removed.

PHYS. Yes, QFT is not an asymptotically free theory, but it is clear that the Higgs spectrum is a phenomenological concept and at the next stage it will be modified.

MATH. Perhaps so. But why was there such disenchantment with QFT at the beginning of the sixties ? A remarkable new variant of the theory had been thought up, namely the Yang-Mills equations, and almost nobody wanted to study them ?

PHYS. We can mention at least two factors, but which of them is the dominant one is difficult to say. At the beginning of the fifties, the theory of the pseudo-scalar interaction of π-mesons and nucleons occupied an enormous amount of space in the "disciplinary matrix" of elementary particle theory. There was a good renormalizable theory. Strange particles tended not to be taken note of since, in practice, it was considered that they provided a small contribution to the interactions of π-mesons and nucleons [15]. The problems that theoreticians tried to solve were of the following type: to calculate in the πN-theory the πN-scattering, the photoproduction and the nuclear forces. There was no success. In 1955, considerable success was achieved in a new direction: by using the general principles of QFT, the dispersion relations for the amplitudes of processes of πN-scattering type were obtained [16]. The amplitudes are quantities that in a certain sense are directly observable; they can be directly determined from the experimental data: the relations were checked by experiment and verified. The obvious interpretation would be that the field theory is valid, but the problem is to find the correct version. However, as is often the case, some completely different programme emerged. Remember atomistics, thermodynamics and Mach. If one takes all the particles participating in strong interactions (they are now called hadrons, but at that time this terminology did not exist), collect together all the amplitudes of the interaction between them, and write down all the dispersion relations, then one obtains a complete system of equations. One might require some further new conjecture or postulate, but in relation to the amplitudes themselves. Such a programme was called a programme of nuclear democracy.

In the sixties, the quark model was already being discussed and there were, so to speak, two concurrent programmes: the programme of nuclear democracy that I have just described [17] and the programme of composite models of hadrons, that is, hadrons consisting of something or other (for example, quarks) [18].

MATH. It is well known that in QED, say, one can obtain by dispersion methods precisely the same results as are given by the Feynman diagrams [19]. How could it be hoped that something else would be obtained in the theory of strong interactions ?

PHYS. If you take a look at the writings of the discussions of the sixties, you will see that many people said as much, while others hoped somehow that the change in QFT had already brought about a new theory [20]. It's always nice to think that we are doing something completely new.

MATH. All this has disappeared, hasn't it ?

PHYS. It's hard to say. In a certain sense, there remained or arose assertions that the Feynman diagrams could be written down for any objects. Neither the π-meson nor the nucleon is elementary, but if nucleons are at a large distance from each other, then they exchange a π-meson, and this can be calculated by simple use of the old Feynman diagrams where the πN-constant is obtained from the Goldberger dispersion relations. If the amplitude of the exchange of a single π-meson falls away as exp $(-mcr/h)$, then that of two falls away as exp $(-2mcr/h)$. Here, perturbation theory does not enter into it. The understanding of this kind of thing came from dispersion methods. The further development of phenomenological descriptions of this kind is an interesting story which is still unfolding.

MATH. You spoke of two reasons for the loss of interest in field theory. What was the second reason ?

PHYS. The second, was perhaps the discovery of the zero charge in electrodynamics. In 1954, Landau and Pomeranchuk produced an argument in favour of the suggestion that the behaviour of the renormalization of a charge, in accordance with QED is such that however large the charge might be at small distances, it would be precisely zero at large distances. That is, QED, understood as a purely local theory, is in fact a theory without interaction. Landau and Pomeranchuk's arguments appear to be fairly convincing even now [21].

MATH. I don't understand all this history. Plausible arguments are all right if you apply them to the real world and to a correct theory in order to get at the answer. If on the other hand you hope to prove that local QED does not describe real electrodynamics, then it would seem that you need a rigorous proof.

PHYS. Or a plausible argument that we should stop tinkering with QED ?

MATH. Somehow or other, these small distances in QED that we have been talking about are uninterestingly small. They were unattainable experimentally. So how could Landau and Pomeranchuk's assertion be brought to bear on anything ?

PHYS. If it is carried over to the theory of strong interactions, then a catastrophe occurs immediately at the distance $\hbar/m_\pi c$. This is how the statement arose that field theory does not describe strong interactions [22]. To what extent this line of argument and the programme of nuclear democracy supported each other is not clear.

MATH. As you know, apparently there was also a programme of Heisenberg's. He considered elementary particles to be pure "platonic forms"; this is, as it were, the anti-Democritus situation. What did he actually have in mind ? You know what he said: "In modern quantum theory, there can hardly be any doubt that the elementary particles are in the final reckoning mathematical forms, only enormously more complicated and abstract in nature" (than the platonic polyhedra) [23].

PHYS. In the articles and books of Heisenberg there are a lot of statements of this sort in his last twenty years. If we analyse the context in which they were written, it is clear right away that we are dealing with several things. Firstly, that in contrast to the atoms of Democritus, elementary particles are not permanent; they are created and they disappear, for example, in the reaction $\gamma+\gamma\rightarrow e^+ + e^-$. This is clearly contained in our paradigm. This is trivial in a certain sense. Particles are not the primitive objects of the theory, which are the oscillators of Bose or Fermi fields. Particles are excitations of them. (More precisely, for Bose fields, their lowest excitations.) The best answer to the question why a newly born positron or electron is always the same that we can currently give, is that the equations for the Fermi field of the electron are always the same. Of course, this is a formal answer; it is in essence a statement of a fact embedded in QFT. It is a long way from that assertion to the statement that the electron is a pure form.

There is a further aspect. In QED one needs to distinguish between the original particle of the equations, "the naked electron" and the real electron which, as follows from the theory, is a superposition of states $e^- \rangle + e^-\gamma\rangle + e^- e^+ e^-\rangle + \ldots$ and is a complex system. Then there arises the question: why cannot the energy or mass of such a system be changed continuously. The answer that suggests itself is that, according to quantum mechanics, the levels of any system are discrete and the first level is a special case. The electron has no excited states.

MATH. In QED we use perturbation theory: does this argument have any meaning then ?

PHYS. Maybe yes, maybe no. Hadrons, which are connected states of quarks, always or almost always have several levels. So that the argument is meaningful here. When protons are created, you always obtain the same protons because you always create the same quarks and because there exist only discrete levels in the system *uud*. The excited proton is already a different particle, also with a definite mass. Once again the formal restrictions of quantum mechanics play an important role.

PHIL. Didn't Heisenberg also mean that the objects became pure forms because they became devoid of ordinary material qualities such as colour, volume, hardness ?

PHYS. They were devoid of colour even for Democritus. Of course, as far as the development of the theory is concerned, elementary particles have become all the more abstract: today we can only talk about them in the language of mathematics. Even so, Heisenberg was no more able to do without "objects" than Democritus; only then, there were individual hard bodies, while now we have quantized fields.

MATH. But all the same, surely Heisenberg had a proper alternative programme. This has been written about in books, for example, in the biography of him written by Hermann.

PHYS. Indeed there was a programme. He hoped to obtain all the particles, leptons, hadrons, γ-quanta as bound states of a non-linear spinor field with self-action [21]. Essentially this was one of the variants of the composite model, but it was proposed that all or practically all the internal quantum numbers obtained would themselves come out as a matter of course. In such a theory the "forms" would have been larger and "materials" smaller than in the modern version in which a large collection of "primitive qualities", such as sorts, colours and spins are introduced. Of course, Heisenberg's programme was also within the framework of QFT; the statement that "elementary particles are pure forms", if this refers to this programme, is no more than a metaphor. Nothing came of this programme.

PHIL. Won't it be revived one day ?

MATH. Programmes are practically eternal.

Dialogue 5

MATH. Yes, programmes are practically everlasting. Perhaps the most beautiful physical theory of recent time that I know of is supergravity. Here, both the programme of the unified theory of gravitation and electromagnetic interaction, which Einstein was preoccupied with throughout the second half of his life, and Heisenberg's programme with his belief in the fundamental nature of the spin 1/2, are brought to life. But look at the form it takes ! Certainly neither Einstein nor Heisenberg would have accepted this theory ...

PHYS. Supergravity is not a theory. It is an unformed plan of a theory which so far does not exist [1].

EXP. That is how it always is with theorists. Just a lot of endless altercations.

PHIL. I don't know. Isn't it just that Math and Phys use the word "theory" to denote different things ?

MATH. Come to that, we have different restrictions on the texts of admissible theories. For my part, a theory must be syntactically organised in a clear cut fashion; mathematicians are accustomed to having strict rules for the introduction of concepts, and requiring accuracy of deductive inferences and the like. Of prime importance for Phys is an extra-mathematical semantics which is the operational aspect of comparing theory with phenomena.

PHYS. As I see it, a theory is a list of postulates enabling one to describe some fragment of reality. In fact, this is most often a set of equations describing the properties of certain objects. A theory exists if the rules are sufficiently clear and enable one to arrive at results which can be compared with experiment. If the results concur, I then say that the theory is true or correct and that the objects introduced in its construction really exist. With regard to "the existence of a theory", this is more of a jargon. The theorist can write down equations but cannot know even a conjectural interpretation; that is, it is not known what the theory proper is predicting; I then say that the theory does not exist.

MATH. It seems to me that you are oversimplifying the situation. For example, Onsager's theory of phase transitions in the two-dimensional Ising model undoubtedly exists, even though no kind of "reality" is being described. Or take the model theory of a gas made out of ideal hard spheres. This again is a perfectly good theory, although it only approximately describes a real gas. I think that the word "theory" describes a whole family of meanings, as in Wittgenstein's famous example.

PHYS. Yes, but in any case, one needs to draw a distinction between theories for which at least an internal semantics is defined, and formulae that are simply written down on paper. I think that with supergravity, even the internal semantics is bad. And when I speak of existing theories, I am not thinking of the model theories or playthings such as the two-dimensional Ising lattice, but genuine theories such as celestial mechanics which describe reality as it is.

MATH. All theories contain an element of idealization and, therefore, a modelling element. In celestial mechanics, the planets are ideally hard spheres or, more often, even material points, whereas in fact this is not the case. All theories, even the best ones, are model theories; it is a question of the degree of accuracy. But in any case , in QED theory, it is possible to calculate observable quantities but in QCD, nobody has succeeded in obtaining from the theory the ordinary hadrons.

PHYS. There are reasonable conjectures about how they could be obtained. Although the problem of obtaining the properties of ordinary hadrons from the theory has, to be sure, not been solved, this difficulty can be overcome, and there are many results obtained in QCD that have been compared with experiment and have given good agreement [2].

PHIL. So will a complete theory of elementary particles be constructed or not ?

PHYS. In principle, we can say that one day when all the interaction theories have been unified, a simple theory will arise containing a very small number of parameters and enabling one to calculate all the properties of the elementary particles. For the moment when, say, QFD contains more than 20 independent parameters we are of course awfully far from this.

MATH. As I see it, it is not a question of parameters. It seems that Phys regards a theory as given, in his sense, when the Lagrangian has been written down. In my sense, this is precisely the point when the theory has been brought to a conclusion. It then becomes an art, a black magic if you like, rather than a theory. Officially this is called "second quantization", "functional integration", "perturbation theory". The bits of mathematics that I am in a position to glean from the text books and preprints promise something beautiful and important, but do not form part of some definitive picture. QCD does not have the power to explain the fundamental phenomenon in strong interactions, namely, colour confinement, but forces us to reject one of the basic dogmas of QFT, namely the completeness of the space of free states of particles. Yes and what about the free states for the non-linear gauge theory where even the quadratic part of the Lagrangian cannot be selected in invariant fashion, and the equations of motion do not satisfy the superposition principle. But let us assume that we can leave this as a problem for the theorists and we say to ourselves: Aha, in the non-abelian gauge theory there has to be confinement. But this is not so; the electro-weak theory with the group $SU(2) \otimes U(1)$ does not want confinement, the symmetry is broken down by the Higgs fields. These Higgs fields describe phenomenologically something unknown; this is clear even at the present level of understanding. But what are we expecting of the non-abelian groups in the unified theories ? Is phase confinement obligatory for them ? Who knows ? We cannot deduce this from the Lagrangian. So we see that 20 constants are a mere nothing. In QCD the Heisenberg equations are non-linear. But the equation for the state vector is linear all the same. What is new is the difference between the spectrum of the free states and the spectrum of the free part of the Lagrangian and there's no need to give up any of the dogmas. With regard to confinement, there isn't any when there are Higgs fields; but we do not know when it is obligatory.

EXP. Yes, I have always thought that the theorists are wasting their time. So what is this elementary particle theory that you are involved with?

PHYS. Here the word "theory" is being used in a different sense. Here a theory is "the totality of our knowledge about elementary particles". The word "theory" is generally used in the most diverse meanings. For example, in the 60's one talked about "the Regge theory of strong interactions". But this was properly only a programme, a hope. There has never been a clear list of postulates, rules of deduction and all the rest of it; and no high-precision numbers either.

PHIL. But I have often seen books called "Quantum field theory" ! Surely some sort of unified field theory is set out in them describing all the elementary particles. [3]

PHYS. Well, not quite ... In any case, we now have two theories of interaction of elementary particles, QCD and QFD, the validity of which is empirically based; they describe behaviour of actually existing leptons, quarks, photons, gluons ...

MATH. I haven't yet finished my tirade (*to Phys.*). Let us go on a bit. Although you say that if a theory in your sense is supported by experiment then the objects named in it exist, this is a somewhat illusory existence. A photon is rather an "event". But what is it outside the act of emission or absorption ? Is the photon a potential possibility of such an act?

PHYS. If you don't like quantum theory, then I am afraid I can't help you.

MATH. But that's not all there is to it. Let's say a photon is scattered by a photon; then after the scattering there is no difference between one photon and the other. And you know perfectly well that n-photons in a state with a given momentum and polarization are generally completely merged, as it were; in Bose statistics the weight of such a state is equal to unity.

PHYS. That is right. But there is no problem here. There is a perfectly clear complete theory, namely, quantum electrodynamics. This theory is, so to speak, a realization of the general scheme of quantum field theory. The theory is perfectly clearly stated. An electromagnetic field is a collection of oscillators with numbers i and an application of the general principles of quantum mechanics to this system leads to the fact that each of them has energy $E_i = N_i h\omega$ where N is the number of the excitation level; the states with $N_i = 1$ which we most often have to deal with, correspond to light quanta. It is known how to describe the interaction of this system with electrons and this can be clearly and unambiguously calculated. Thus there are no problems here. Of course, because of the unexpected and unusual nature of quantum mechanics, the inexperienced person may stumble across an apparent paradox or contradiction, but these can always be resolved; this is not a difficult matter for the good expert in the theory.

PHIL. But there are divergences in QED are there not ?

PHYS. The interactions of electrons and photons are usually calculated in perturbation theory [4]. Here the simplest act is the emission of a photon by an electron. (*Goes to the blackboard and draws the following diagram.*)

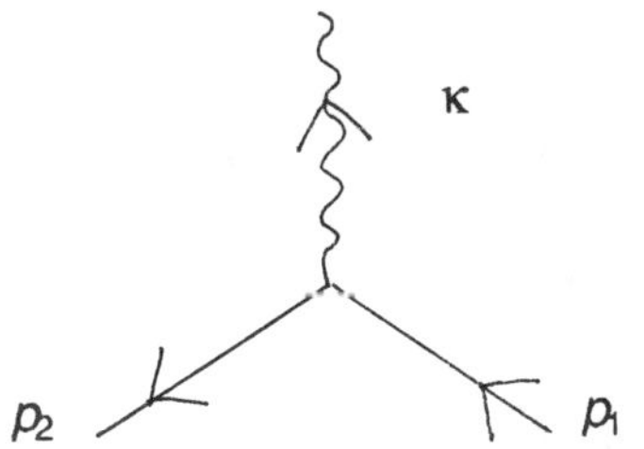

$$p_2 = p_1 - \kappa_1$$

Fig. 1

An electron with four-dimensional momentum p_1 emits a photon with momentum κ and goes into a state with momentum $p_2 = p_1 - \kappa$. The amplitude of such a process is defined and is proportional to a constant $\sqrt{\alpha}$. More complicated processes are constructed from simpler ones. For example, the scattering of a photon by an electron is described by two possible diagrams.

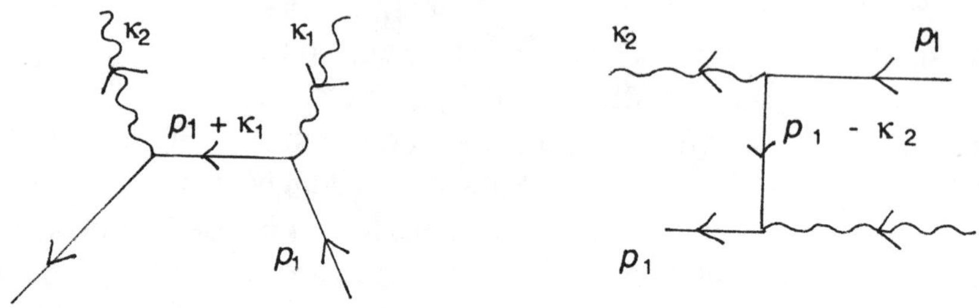

Fig. 2

MATH. You should say that for a free electron, the process $e+\gamma\leftarrow e$ is impossible, and that in your diagrams the electron between the initial and final acts is not real but virtual.

PHYS. Of course, but in fact what I am describing does not in general have to be taken literally. This graphic description of a calculation scheme was invented by Feynman. In fact, it is possible to perform the calculations by other methods. In all cases we represent the state of the interacting fields as a sum of states of non-interacting fields. The difference between one and the other must somehow show up. In the language of particles, we say that particles in the intermediate states are "virtual". This is in fact an imperfect form of the expression of the fact that in reality we are dealing with non-free states. I don't think this is important right now. In the lowest approximation with respect to α, there are no divergences for any process, and all questions of identity of particles and the relation between waves and particles and all other questions of the this kind can be completely solved. The question as to what to do with more complex diagrams where divergences arise is quite another matter but even there, calculations are possible.

EXP. All this is typical abstruse talk of theoreticians. Of course, elementary particles are just particles. I see them very well in the laboratory apparatus. Unfortunately I don't have any more time to listen; I have to go to CERN where an experiment of mine is in progress. (*Leaves*).

PHIL. In the final event you must admit that his experiments are the ultimate reality. But should I then believe him ?

PHYS. You would be mistaken. It is simply that his apparatus was so constructed that he got used to dealing with geometrical optics.

MATH. All the same, you talk about photons as though they were particles, but in fact they can't be localized more accurately than the wave length of λ. Just think of a radio wave of wavelength 1 km. What sort of a particle is that ?

PHYS. The primary concept of the theory is the field, not the particle. One can only talk about particles approximately in the limit when one can use geometrical optics and when we deal with single-particle excitations. In this case the whole system of concepts such as trajectories, particle momenta, their energy, and so on works. There is no reason to be astonished that geometrical optics does not work when it does not have to. (*to Phil*) Nature around us functions on a large scale by electromagnetic (or gravitational) interactions, until we broach the nuclei.

MATH. This is very amusing. Nature teases us, as it were. By itself, the division of natural phenomena into layers that are largely isolated and where one or two simple interactions dominate has, it would seem, been specially designed so that we can study theoretical physics gradually. Look how wonderful it is: on the galactic and planetary scales, gravitation suffices in order to explain the motion of the stars and the planets; and in nature around us, when we study the enormous diversity of atomic phenomena, just the electromagnetic interactions suffice. To explain how and why the stars shine, we need to take into account gravitation, the weak, the electromagnetic, and the strong interactions. It is only at the very early stages of the evolution of the universe that one begins, apparently, to take into account the quite short-range forces predicted by GUT. But on the other hand, nature is somehow indifferent to theoretical rubrics. At the beginning of this century one spoke about atomic theory, which was supposed to explain the observable properties of matter; now this notion has almost disappeared, and not by chance [5]. At the end of the forties the notion of atomic theory was replaced by that of "atomics physics", which in turn has now completely disappeared, or so it would seem. And this too is not by chance. When studying "nature around us", we sometimes make use of non-relativistic quantum mechanics and non-relativistic Coulomb interactions, sometimes rudimentary elements of the quantum theory of electromagnetic fields, when the interaction of light with matter is being described and sometimes when, say, we are analysing in detail sound in gases, we simply use the Boltzman equation, that is, classical mechanics. Sheer eclecticism.

PHYS. You have somewhat overstated the vagueness of the situation. Actually we could act like Fermi as in Heitler's book [6]. We start with the total Lagrangian of electrodynamics, we then separate off the Coulomb and then study the free plane waves, that is the photons. And this will then be our world of nature.

MATH. I haven't managed to make my main point. Ordinary processes do not change the nuclei but, of course, nature around us is a consequence of their existence. Of course, for the majority of processes you consider them entirely à la Boskovich, as Coulomb centres with mass Am_p and charge Ze. But from a more elevated point of view the troubles begin; nuclei consist of protons and neutrons, and these are extended objects of dimension of the order 10^{-13} cm. Their properties, such as the distribution of charge and mass, depend on the strong interaction, if you want the level of precision that you may have to take into account; the situation then opens out and you are already entering the world where not only QED rules but also QCD. The theorists who like tidy structures prefer to speak of the quantum electrodynamics of photons and electrons. Here for quite a long time we can calculate the higher approximations in α without running into difficulties.

PHIL. What do you mean by "quite a long time" ?

MATH. At very small distances one cannot perform renormalizations without contradictions.

PHIL. Why ?

MATH. This is already another story.

PHYS. In practice, at some moment, other interactions have to be taken into account; they can in the end remove the difficulties.

MATH. I don't like the fact that all the time we have to appeal to some sort of deus ex machina which has to retrieve the situation somehow.

PHYS. Surely your colleagues of the Eudoxus era are to blame for this. They have foisted on us too simple a model of the continuum. Hence the divergence and all the rest. If you could suggest to me something a little better, I would probably suggest to you a non-contradictory theory. However we are already succeeding some of the time, it would seem; QCD does not contain contradictions at small distances.

MATH. I don't agree.

PHYS. But it's obvious. Remember that the charge falls away at small distances and all the problems disappear.

MATH. That's true, but the theory as understood is clearly inadequately formulated. We started off with a point interaction and now, it would seem, a point does not have an interaction.

PHYS. I admit that we do not have an adequate language for the description of our calculations, but essentially I think that QCD is the correct theory for strong interactions and the problem of small distances does not occur there.

PHIL. It seems to me that you and Math are saying that the mathematical apparatus acts correctly but your ideas about its meaning are not totally correct.

MATH. Or totally incorrect. Consider Maxwell's time. You have a microscopic electrodynamics which at least for macroscopic bodies is correct just as it is now, and one can say that it is completely true. On the other hand, Maxwell thinks that this is the Lagrangian mechanics of the Newtonian ether. Apart from this, the theory contains roughly phenomenological components. Apart from the fields E, H there are also the fields D and B and, finally, the equation for the current $\mathbf{j} = \sigma\mathbf{E}$. All this will be got rid of; the ether will disappear, the banished electrons will return and in 1906 Planck will write down the action for the point charges and electromagnetic field that now feature in the second volume of Landau and Lifshits.

PHYS. I willingly admit that the Higgs fields of the electro-weak theory are also a very crude phenomenology. But if Maxwell had not taken his ether seriously, he would not have proposed the Michelson experiment. It isn't good for theorists to call their theories into question. Development proceeds by trying to expand them as much as possible. It is only when we stretch them to the limit that we break them.

PHIL. I think it is time to stop while everything in the room is still in one piece.

Dialogue 6

PHIL. At home I was looking at the dictionaries: there also, the word "theory" has various meanings. (*to Phys*) When you said that the theory of supergravity does not exist, did you have in mind the example "Theory 3" in the sense of Amer. Her. Dict. ? [1]

PHYS. Primarily I had in mind the project "supergravity extended to $N = 8$". Math would say that there is a large family of geometric theories to which the word "supergravity" is applicable but which differ from one another in the number of anticommuting coordinates in the superspace, the choice of the Lagrangian etc. As I see it, in order to describe all particles, one needs a maximally extended geometry. But apart from this, I am convinced that the system of assumptions embedded in this project will merely produce nonsense when the rules of procedure have been worked out. Stephen Hawking, who believes in supergravity, probably has in mind something like "Theory 3". In any case, I already explained last time my own prerequisites for a physical theory. I think that in recent years this word has come to be used in a somewhat romantic fashion.

MATH. Are you saying that $N = 8$ supergravity is physically uninterpretable ?

PHYS. More simply I mean that when one gets down to calculating the cross sections of processes using this Lagrangian, one obtains only infinities.

MATH. But don't forget that in your favourite QED, if you simply calculate the series of the perturbation theories, then again all the terms will be infinite.

PHYS. You know perfectly well that everything is finite after renormalization, which allows us to calculate any observable in practice to within any number of decimal places.

MATH. The mass and the charge of the electron cannot be calculated.

PHYS. And there's no need to; these are parameters of the theory in terms of which all the others need to be expressed; for example, the cross section of the scattering of light by an electron.

PHIL. Is the "cross section" an "observable" of quantum mechanics, then?

PHYS. (*uneasily*) It would be nice to be able to say yes, but in fact this is not the case. An "observable" is something else here, namely, the "angle of scattering θ"; while the "cross section of the scattering at an angle θ" is the probability of measuring a photon in the corresponding state.

PHIL. Aren't these the probabilities that Exp measures ?

MATH. Sometimes, yes.

PHIL. And what else ?

MATH. Generally speaking, he is more fond of measuring "the parameters of the theory".

PHIL. I should like to see how all this looks; I still remember my school algebra.

PHYS. We need a blackboard.

MATH. (*Gets up and brings the blackboard on castors from the other room.*)

PHYS. (*Writes down the formula*)

$$\sigma = \frac{8\pi}{3}\left(\frac{e^2}{mc^2}\right)^2$$

PHIL. Is σ the "cross section" ?

PHYS. Yes.

PHIL. But where is "the angle θ"?

MATH. He has written down "the total cross section". This is the probability of the scattering summed over all angles.

PHIL. I should like to see the "scattering angle".

PHYS. (*With annoyance*) I don't remember, but if you really want to know, just give me a few seconds. (*Mutters discontentedly under his breath and starts to write something in the right hand top corner.*) Ah, here it is for unpolarized light

$$d\sigma = 4\pi\left(\frac{e^2}{mc^2}\right)^2 \cos^2\theta \sin\theta \, d\theta$$

MATH. You haven't written down the quantum formula for him but the Thomson cross section !

PHYS. That's right, but this is the QED result as $\omega \to 0$. (*Turning to Phil.*) Here e is the charge of the electron, m is its mass, π is pi and c is the speed of light.

PHIL. And is c also a "parameter of the theory" ?

MATH. No, it is a "universal constant".

PHYS. What's the difference ?

MATH. It is connected with the Poincaré group and is universal. It enters into all theories, not just QED. Properly speaking, it's best to choose the units of measurement so that it is equal to 1.

PHIL. Are there any other universal constants ?

PHYS. Yes, $\hbar$.

MATH. What about the gravitational constant ?

PHYS. I don't know (*uncomfortably*). Perhaps the difference is one of convention, although I have my own thoughts about it.

MATH. When you wrote $d\sigma$, you were writing from memory. Is the deduction of this from the Feynman diagrams complicated ?

PHYS. Just a moment. (*Writes.*)

PHIL. But where is the "series" ?

MATH. He has written the first term.

PHIL. And what do the others look like ?

PHYS. They correspond to the diagrams ... (*Draws a diagram.*)

PHIL. And what about the formula ?

PHYS. What's it to you ? After all you're not Exp and you're not getting ready to measure something.

MATH. They have computer programmes that can calculate these diagrams on their own.

PHYS. Well, not quite, they can't perform renormalizations.

MATH. Well they can write that down as well, but it's not essential. All the same, the "axiomatics" of your QED are contradictory, and in the logical scheme, you don't know what you are doing.

PHYS. I don't have to have an ideal theory. I know from physical considerations that the processes that I have calculated and over which I sum occur at certain distances. At very small distances QED would not work as well, and I perform my calculations "right up to these distances". If I calculate the mass of the electron and its charge, then I obtain (in the next term of the series in e_0)

$$m = m_0(r_0) + \frac{g}{2\pi} e_0^2(r_0) \ln \frac{1}{m_0 r_0}$$

$$e^2 = e_0^2(r_0) - \frac{2}{3\pi} e_0^2(r_0) \ln \frac{1}{m_0 r_0}$$

Here $m_0(r_0)$, $e_0(r_0)$ are the charge and mass inside the "sphere" with radius r_0. If we now calculate the cross section in terms of m_0 and e_0, then $\ln(1/m_0 r_0)$ enters into the expression, but the observable mass and charge are m and e. By expressing σ in terms of m, e, one obtains a quantity not depending on r_0. This procedure is called renormalization.

PHIL. Can I have a look at the answer ?

PHYS. The calculations are lengthy, but they can be extracted from a book. (*To MATH.*) Do you have anything on QED ?

MATH. There's Akhiezer and Berestetskii. Here you are.

PHYS. (*Writes down the formula.*)

PHIL. And how long did it take your predecessors to sort this out ?

PHYS. About thirty years.

MATH. All the same, you have introduced a few phantoms into your renormalizations; you say that in the vacuum, e^+e^- pairs are created, then the electron carried into the vacuum attracts these particles or repels them, and from this there emerges the charge of the electron depending on the radius r; but all this reminds me very much of the cog wheels and constructions that Maxwell invested in the ether. But it's not like this at all, and your Dirac agrees with me.

PHYS. I know; but it's not like that. Let's forget about the closed QED of photons and electrons for the moment and place a nucleus in the vacuum (its dimensions are 10^{-13} cm. which as an approximation we can neglect). Then the nucleus can also pull apart the vacuum pairs e^+e^-, or as they say, it will polarize the vacuum. The typical dimension of the charge cloud formed around it is $\hbar/(m_e c) \sim 10^{-11}$ cm. As a result the field of the nucleus is somewhat different from the Coulomb field. This can be calculated and observed. For this, one needs to place a μ-meson close to the nucleus. The μ-meson is heavy; the radius of such a μ-meson-atom is $\hbar/(m_\mu c Z\alpha) \sim 10^{-12}$ cm. for $Z \sim 10$. Such a μ-meson sits directly in the cloud and measures its density [2].

I am quite confident that up to distances of 10^{-16} cm., QED describes reality, so to speak, in one to one fashion. Everything that can be calculated can, in principle, be measured.

MATH. Where does the 10^{-16} cm. come from ?

PHYS. From measurements where QED was tested . In order to get to such small distances, these have been done in fact with leptons.

PHIL. Why were leptons singled out ?

PHYS. They have no structure. They are "point particles".

MATH. But you yourself have explained that in QED a certain amount of pruning has to be done; in other words there is some structure somewhere.

PHYS. Be that as it may, they have no structure up to distances of 10^{-16} . I can do some pruning at this radius, and then the unified Weinberg-Glashow-Salam theory will work.

PHIL. So you're saying that different theories apply at different distances?

MATH. Yes, that's how they have been constructed. I spoke about this the last time, and it greatly simplifies life.

PHYS. This is fairly simple. Here we have different classes of particles and different types of interaction. Some act at small distances and some at large distances.

PHIL. And presumably the gravitational ones act at the largest distances of all ?

PHYS. Yes; the electromagnetic interactions as well. And the "theories" describe specific types of interaction.

PHIL. If I understand you aright, as soon as a "theory" is formulated, the algorithms for calculating the observables are fairly simple.

PHYS. Yes, but they may require a large amount of work. The most powerful computers can hardly cope with calculating α^4 orders for QED. But the most important thing of all is that the procedure of renormalization cannot be carried out in all the theories. There are "non-renormalizable" theories. In general they are nonsensical.

PHIL. What is this quantity α that you were referring to ?

PHYS. The series in the form that results in QED are obtained are series in the parameter $\alpha = e^2/(\hbar c) = 1/137$. Since it is small, everything can be easily calculated.

MATH. Do you think that supergravity is non-renormalizable ?

PHYS. Yes, of course.

MATH. But, maybe, there is some as yet unknown method of extracting the final answers from the theories ? After all, in the 30's, nobody knew that QED was renormalizable. According to your logic, it should have been rejected.

PHYS. We now understand everything better than in the 30's. A theory is either renormalizable or nonsensical.

MATH. But why shouldn't it be finite in all the orders after the Born order? Such examples are already known. Or the theory may be summable and finite after summation.

PHIL. I can hardly make any of this out, but all the same I should like to know the difference between quantum mechanics and quantum field theory. Also I have not yet grasped how many interactions there are and how many theories. For instance, you have just said "there are examples ..." . Examples of what ? And what are these "sets of phenomena" ?

MATH. "Quantum field theory" is not a theory in the sense of your dictionary. It is rather like a grammar, a rule of constructing "theories", a sort of universal language. Within the bounds of this language you can construct different "theories" by choosing different "sets of basic fields", "Lagrangians", "the number of dimensions of the space". At present theorists know two

Lagrangians: one is the Lagrangian of the electro-weak interaction, the other is that of the strong interaction; and it is thought that these Lagrangians describe these interactions. The rest of the theories are, for the moment, toys or models. The toys are for testing and studying the properties of the various versions. They can be studied without concerning oneself about whether they describe anything.

PHIL. But why haven't you said anything about the Lagrangian of the gravitational interaction ?

PHYS. It isn't completely clear how to quantize the gravitational field.

MATH. In my view, it is completely unclear.

PHYS. I think that for the weak fields everything is in order and the Born approximation can be regarded as being correct.

MATH. In what sense correct ?

PHYS. If it were possible to scatter gravitons by electrons ...

PHIL. What's stopping Exp from "scattering gravitons" ? After all, he already knows how to scatter photons.

PHYS. It is not possible to make a sufficiently powerful source of gravitons; also there aren't any receivers.

MATH. You are contradicting yourself. Don't forget that the gravitational Lagrangian is not renormalizable.

PHYS. But it is correct in the classical approximation.

MATH. Is non-renormalizability important or not ?

PHYS. Maybe so; if only because it enables me get to the end of a calculation. But then again, maybe not. Perhaps the Einstein Lagrangian needs to be generalized to some other Lagrangian in order that it should be renormalizable.

PHIL. Is this so that the theorists who are involved in elementary particle theory can occupy themselves with writing down the various Lagrangians and working them out ?

PHYS. Yes, either with the object of describing phenomena or in order to learn something about Lagrangians in general. But fields still need to be quantized.

PHIL. And which theory predicts the proton decay ?

MATH. Gauge theory, which unifies the electro-weak and strong Lagrangians.

PHYS. It is not clear whether such a theory exists and to be sure, there is not a unique version but several.

PHIL. Well, does one get non-renormalization problems ?

PHYS. No. But it's not just non-renormalizability. There are other problems as well.

MATH. This is in fact the main trouble with physicists. You can never obtain a list of assumptions from them, or a list of problems either.

PHYS. If we had started with a list of assumptions, we'd still be studying $\mathbf{R}^3$. (*With annoyance*) We are thinking about actual realities.

PHIL. As I recall, Exp said that your theories always turn out to be erroneous.

MATH. It can't be helped. On the road from "theory" in the sense of item 3 of Amer. Her. Dict. to "theory" in the sense of item 1, the casualty rate is very high; some theories die from internal contradictions, others from disagreement with experiment.

PHYS. To finish up with, supergravity with $N = 8$ according to this idea, should unite all the interactions including gravitation. Formally this is non-renormalizable.

Dialogue 7

PHIL. I am already getting used to the vague ways in which theorists such as yourself use the word "theory". But now I should like to return to "paradigm". I have again tried to compare the dictionary entries with your use of the word. Can one say that the paradigm of a given period of development of a given science is a set of theories constructed in an analogous way and apparently competing (*turning to Phys.*) as the candidates for the description of the notorious "reality". The realization of a claim of some specified theory then *exclut la réalisation concomitant des autres* etc. Is that how it is ?

MATH. I think that's roughly the case. The term "paradigm" was successfully introduced into the history of science by T. Kuhn of the U.S.A. in the book "The structure of scientific revolutions" [1]. True there is no particularly clear definition there, but then, you know, it's always like that. The speaker or writer looks for names for phenomena that he has observed (paradigmatic character of the thinking of the researchers of a given period). He borrows the terminology of the linguists and then provides himself with his own definition that perhaps is not quite in accord with his basic intentions; if nevertheless, this is relevant, then the term starts to be used by others. I use it, in fact, in the sense that you have suggested, but in theoretical physics there is a further aspect: in each field of activity, some definite language of description with a reasonably clear grammar and dictionary is operating. The rules of formation of theories belonging to the paradigm of some period are in fact always very rigid, (*to Phys.*) so that you theorists have little freedom in general. So there it is. When I have been speaking about a paradigm I have apparently had this fact in mind and sometimes my paradigm is simply a dictionary with grammatical rules, rather than a "set of theories". I tend to associate the word "paradigm" not only with the theories themselves, that is, with *la parole de la physique théorétique,* but also with the generating mechanism of this language, its *langage* or *langue* in the sense of Saussure [2].

PHYS. It seems to me that Kuhn himself had not been too worried about making this word a term. Somebody has told me that if you were to copy out in succession all the phrases in his book with the word "paradigm", the result would be pretty vague.

MATH. That may be so. But I have already told you the sense in which I use this word and I do think that in general I don't abuse it. Incidentally, there are some similar examples somewhere in Kuhn. But in any case, when we are trying to nail down terminology, we no longer need to study Kuhn's text; instead we must understand the term in its real medium of existence.

PHYS. So in spite of all this, when you use the word "paradigm", what do you mean: "la parole" or "la langue" or "language"? That is, do you mean a set of theories or " a generating principle", or a dictionary and grammar of theories ? And if you mean a generating principle, then where is it localized; in the minds of the theorists ?

MATH. I think that for the moment we can satisfy ourselves with the level of rigour that has already been attained.

PHIL. Theoretical physics is a fairly young science. I have recently read the suggestion that Newton's "Elements" marks its birth date. And the study of science is altogether in its infancy.

MATH. In general this is roughly the case, although this same author elsewhere in his books writes that the first physical theory was Euclid's Geometry [3], and we can agree with him in this respect as well; in particular, this is precisely the opinion of von Neumann who, incidentally, has observed that Newton's "Elements" directly imitate the form of Euclid's "elements" [4]. In fact, Archimedes's statics and Ptolemy's geometrical optics are also entirely "theories". I think that if there had not been the political and ideological catastrophes of latter antiquity, then we would consider Euclid's geometry as the first physical theory but the clocks had been stopped for a long time [5] .

PHYS. The Greeks had no experimental method.

MATH. Have a look at Dorfmann's History of Physics on experiments in optics.

PHYS. Do you think that the absence of the experimental method in Greek science is a myth ?

MATH. There has been a misunderstanding. Between, say, the second and sixteenth centuries there was a stagnation in the European science. In the period up to the 15th or 16th centuries, Aristotle was held in too high regard, and when time came to move forward, it was felt that the dominance of experimental proofs should be emphasized.

PHYS. In fact there occurred a big shift in orientation. To all appearances, the Greek scholars believed that the final truth of any topic could be attained by means of rational argument; in the final result it should become self-evident in the process of discussion.

MATH. And what do you call rational argument ?

PHYS. Well, roughly speaking it's what Lakatos called the method of proofs and refutations [6]. So to speak, the dialectical process leading to a final clarification of concepts.

MATH. It seems to me that modern theorists sometimes indulge in this.

PHYS. Yes, to a large extent; but with them, the rules are changed. In general, they do not hope to obtain a "definitive" theory and are satisfied with provisional theories having a phenomenological status and describing some fragment of reality. The justification of such a theory is not in its logical necessity but in its empirical truth. What is more, the basic premises can be quite strange and not to the liking even of the creator of the theory. Let us say, Newton thoroughly disliked universal attraction acting at a distance but he knew from experiment that it described phenomena. In principle he most likely did not believe that it was just the gravitational action that was the primary fact, but he was prepared to accept it as a method of describing phenomena. In other words, the orientation toward phenomenology appeared.

PHIL. (*not without malice*) So wasn't this Christian humility before nature and her creator ?

PHYS. I can't prove that this hypothesis is false.

MATH. But do you in fact reckon that the current paradigm of QFT is also a phenomenology and that it has to be replaced by another with another grammar and another syntax ? Einstein hoped for a definitive theory.

PHYS. Einstein did not approve of the change from the paradigm of classical field theory to quantum mechanics. He always reckoned that he was directly contemplating divine truth, and it turned out to be nothing like quantum mechanics.

PHIL. Here again you are talking about quantum mechanics rather than quantum field theory; so please, can you state clearly which is which ?

MATH. There are certain subtleties. Strictly speaking there were just two grammars or languages in theoretical physics; the grammar of classical Hamiltonian systems, or the language of Hamiltonian mechanics (or simply classical mechanics) and the grammar, or language, of quantum mechanics. But there are systems with a finite number of degrees of freedom; for example, the planets and the sun, and there are systems of the classical electromagnetic field type. The latter are also Hamiltonian systems but with an infinite number of degrees of freedom. It is the same with quantum mechanics. Its language can be applied to a finite number of degrees of freedom, say, an atom with n electrons, as well as to a system with an infinite number of degrees of freedom, say, an electromagnetic field. In the first example, we are dealing with "non-relativistic" quantum mechanics and in the second, with quantum field theory.

PHIL. And do we have relativism here ?

MATH. For a physical theory, its invariance group is of the utmost importance. Experience has taught us that there is one universal invariance group, namely, the Poincaré group "*P*". This, as they say, is the invariance group of space-time of the special theory of relativity. If we wish to construct theories that are invariant with respect to *P*, or relativistic for short, then we have to consider field theories, either classical or quantum.

PHIL. What do you mean by "theories that are invariant ... "?

PHYS. This means that the equations of the theory are taken into themselves under the group transformations. Let me clarify this with an analogy. The equation of the circle with radius r and centre at the origin $x^2+y^2 = r^2$ is taken into itself (that is, it preserves its form) under the transformation (or change of variables) $x' = \cos\theta \,.\, x+\sin\theta \,.\, y$, $y' = -\sin\theta \,.\, x+\cos\theta \,.\, y$.

PHIL. So are the equations of your "field theory" similar to this equation?

PHYS. Not quite. The equation of the circle describes a concrete object while the equation of field theory is a whole infinite number of situations that in general can be described in the language of the theory.

PHIL. Could you clarify this ?

MATH. Yes. The equations are constructed in such a way that one needs to "define the situation" at the instant $t = 0$, while the equations predict its evolution. In fact, in the classical theory, any situations that can be described in the language of the theory are admissible, while in the quantum theories this is not quite true; the equations restrict the admissible class of initial situations. In a certain sense this can also be taken into account in the language: one does not try to assign values to those classical quantities that are represented by non-commuting quantum operators.

PHYS. I should like to return to the notion of paradigm. It seems to me that whatever the true intentions of Kuhn were, he understood the term paradigm in a somewhat more interesting way. For him it is rather some sort of sociological concept, a norm accepted by some group of investigators: so to speak, that which is considered correct and that which is not. And this is accentuated by the group and, maybe, irrational aspect of "scientific behaviour". Something is taken to be true not because some person knows rational arguments supporting it, but simply because this is what his group does.

MATH. That may be so. But surely this element is always present in the development of any science.

PHYS. Kuhn has this other interesting notion of "disciplinary matrix". Roughly speaking it contains components such as: "symbolic generalizations", "metaphysical paradigms", "values", "samples". The first two are roughly the "fundamental concepts of the theory" and "ideas about reality", while the other two are rather obvious.

PHIL. And what are all these "values" ?

PHYS. Kuhn is not a very clear thinker. I would imagine that "values" are the prejudices of a given group or a given investigator. For example, Einstein supposed that the space-time description of systems should of necessity be preserved, and since quantum mechanics did not satisfy his prejudices concerning how such a description ought to function, he did not accept it. I don't expect that "samples" present any difficulties for you. Let us take QED for instance. This is a "model" for the theory of the electro-weak interaction. It would seem that Kuhn originally understood by "paradigm" what he later proposed calling the "disciplinary matrix". If one understands by paradigm the whole collection of these four components, then this is richer in content than simply the "language of quantum field theory".

MATH. Even mathematical constructions are not easy to completely describe in a formal manner. Perhaps you would like to find an ideally regulated terminology for the history of science and to give an ideally clear model ?

PHYS. Of course not. What is more, I am quite convinced that this is impossible. The creation of science is a form of human activity, and just like any human activity it can least of all be fitted into some rigid framework.

The activities of researchers are much freer and less paradigmatic than the science buffs think. In fact each theorist obtains from the literature he has studied and from his teachers a certain knowledge of the concepts, images and theories already known. And then all this depends on what he intends to do. He may devote himself to areas where the basic principles are already known, say, the theory of rigid bodies [7]. Furthermore he will know that his object consists of atoms with collectivized or non-collectivized electrons and that quantum mechanics is valid here. His normal science is prescribed but within its framework he can achieve remarkable results, for example, he can construct a beautiful theory of superconductivity.

From an entirely different standpoint, the theorist who is concerned with elementary particles just does not know how long his paradigm will hold up (to date, quantum theory). He is fully prepared for a catastrophe which, in principle, he considers desirable; that is to say, it is his "moment of truth" when he sees the hazy shore of a new land, "the new paradigm". Until this comes to pass, he will of course try to apply the old paradigm in order to explain new phenomena; he will do this simply because he has no choice. This may prove successful; the limits of the paradigm may be broader than anticipated.

Furthermore, the practical theorist, especially at the early stages of broaching a new region of investigation, will act very informally; his images will be considerably less tied down by paradigms than might have been suggested a priori; he will be prepared to mix all his paradigms, the old ones, the new ones and his personal ones, if only to achieve success.

PHIL. I have already heard some of the details. But I should somehow like to find out about quantum field theory more systematically.

MATH. I have tried to write a small popular survey of quantum field theory. Would you like to try and read it ?

PHIL. Certainly. But I would like to hear again what you yourself call a paradigm; forget about Kuhn !

MATH. Well, roughly speaking I would call a paradigm of a given period of development of science the deepest layer of the fundamental concepts and grammatical rules that are used in the *Theorienbildung*. At present it would seem that for theoretical physics these are the concepts and rules of quantum field theory.

PHIL. Are the fields defined in space-time ?

PHYS. Yes. Math. would say in the space $\mathbf{R}^4$.

MATH. In fact, the theorists abandoned $\mathbf{R}^4$ some time ago. They work in superspaces and in spaces of arbitrary dimension.

PHYS. These are all models or speculations; we live in the space $\mathbf{R}^4$.

MATH. In fact, if you look at Einstein's general theory of relativity, you will see that this is no longer true.

PHYS. Locally, this is $\mathbf{R}^4$ just the same. In any case, as long as we are dealing with theories whose relationship with reality has been established, the only thing that we use for them is $\mathbf{R}^4$. Multidimensional spaces and superspaces are merely uninterpretable wild fancies.

PHIL. I see that everything is not so clear with this paradigm.

MATH. You're right, it isn't. Everything that we have been considering is field theory. There are the extension of types of variables, dimensions and topology of spaces, these are fairly simple experiments with language. This is rather an illustration of the paradigmatic thinking of each period. In any event, we must admit that in going from the classical to the quantum paradigm some tremendously deeper change has occurred in the grammar and the language than what is now being attempted in order to extend the framework of the theory.

PHIL. And what do you call the classical paradigm ? Is it the language of classical mechanics ?

MATH. If we make a serious analysis, then yes. In the final reckoning, in the 19th century there was no more fundamental system of concepts than the language of the mechanics of central forces.

PHIL. But in fact, in the 19th century Maxwell's equations were already well known, and this can be considered to be a new paradigm, namely, the paradigm of classical field theory.

MATH. Ah, but it was believed that behind the Maxwell equations stood the mechanics of the ether.

PHYS. Certainly. But all this shows is that it is possible to go imperceptibly from one paradigm to another.

PHIL. Well, let's have one more cursory look at the history. How many were there of such basic paradigms in all ? In the end, we can only come to an understanding about classical field theory. Suppose that there are three or four: classical mechanics of central forces, classical field theory, the quantum mechanics of non-relativistic systems and the quantum mechanics of fields. Can you think of anything else ?

MATH. Perhaps so. There was the ancient atomistic theory. At the bottom of it lay the very simple idealization, namely, the notion of an ideally hard solid body. The atomic theory as set out by Lucretius was also a paradigmatic system and a very remarkable one at that [8].

PHYS. Oh come on! There was essentially very little science in it; it was more of a philosophy.

MATH. That's true. There is very little that can be described in this paradigm but anyway, when Maxwell wrote his first article on the kinetic theory of gases, he in fact worked within this paradigm, and quite deliberately so [9].

PHYS. Perhaps so. But of course, transferring to the paradigm of central forces did not present any difficulties for him.

MATH. Of course not, because he knew it well. On the whole, the paradigmatic nature of scientific thinking is almost a linguistic fact; we are limited by the vocabulary and grammar of the scientific language in which everybody around us speaks. Even in our attempts to extend the language we are restricted all the same by the resources, say, of mathematics.

PHIL. So how does the change of language come about ? You say that quantum theory is a completely different language. How did it arise ?

PHYS. Force of experiment caused one to transfer new meanings to the old vocabularies or even to introduce radically new concepts and images such as Heisenberg's "Tableaux" which then turned out to be matrices of operators. Then as the language progressed, we learned about the language of Hilbert space. This already is quite another story.

PHIL. That's all very well, but if you talk about the paradigm that is already in existence, then of course at every period the dictionary and the grammar are rather extensive. How do you see the basic paradigm ?

PHYS. Most specialists in a given area are, perhaps, not too bothered about this question. They take the vocabulary and grammar on trust and concern themselves with concrete problems. But at a deeper level, it would seem to be a question of such things as potential completeness, closedness, reducibility, and absence of contradiction of the description. For instance, at the beginning of the 19th century the theorists finally "knew" that the mechanics of continuous media could be deduced from atomic theory and the mechanics of central forces and they were even preoccupied with such matters; that is, they took the paradigm of mechanics as the basic one. On the other hand, it was the actual language of the mechanics of continuous media that could hardly be accepted as basic, but rather as being deducible from the mechanics of central forces. Helmholtz deduced the law of conservation of energy from the mechanics of central forces [10]. Most likely, it was well understood by all that the mechanics of central forces was, so to speak, a complete closed scheme, that is, if the forces and the masses were prescribed, then the motion could be calculated.

Dialogue 8

PHIL. I read the extensive article by our host but I think that I only understood one thing: the modern theory of elementary particles reduces to certain not too complicated "Lagrangians" which are required for the calculation of the "Feynman diagrams", and (*turning to Math*) that for a man as knowledgeable of the *language* of modern mathematics such as you, all this can be explained in 80 pages. And it seems to me also that you reckon that the mathematical procedures of calculating "cross sections" and "lifetimes", which are "observables" in some sense, are very ill founded as you see it. But after your recent discussion with Phys on "the Thomson scattering cross section" I was left with the impression that after the "Lagrangian" had been written down the rules themselves could be stated quite briefly.

PHYS. That is true. This is what all students do. In principle, you can calculate the "Thomson" cross sections by patiently multiplying out the matrices and "taking the trace" [1].

PHIL. What is a "trace" ?

MATH. By the trace of the matrix A_{ik} I mean $\sum_{i=1}^{d} A_{ii}$, where d is its dimension. It is denoted by $\mathrm{tr}\,A$.

PHIL. And is that all there is to it ?

PHYS. Yes. But if you want a clearer understanding of what the "Feynman rules" are, read his famous booklet "The fundamental interactions" [2].

MATH. Yesterday I tried to calculate Thomson's σ, but I didn't obtain the answer in the book.

PHYS. Well then, you must have made a mistake somewhere.

MATH. I realize this. But how does one find the error in such a rather lengthy calculation ?

PHYS. Consider the passage to the limit as $\omega \to 0$ at each stage of the calculation !

PHIL. I think that you are getting away from the point. Right now I am not interested in the arithmetic but I should like to know your opinion on the soundness of the recipes: is Math right ?

PHYS. Certainly not. Newton and Laplace calculated the motion of the planets long before Poincaré introduced the notion of asymptotic series, and their calculations did not in any way become worse after Poincaré had explained that their series were not convergent but asymptotic. The series of QED describe observable data with tremendous accuracy (up to 10^{-9}). One cannot get away from this fact.

MATH. But you know full well that the theory is not closed. At small distances, the charge increases and even you will have to abandon the calculation of this same $\sigma(\gamma + e \to \gamma + e)$ at sufficiently large energies and momentum transfers.

PHYS. Well, basically I don't like these series, but with their help the results of the first approximation can be considerably improved.

MATH. But once the theory is not logically closed, then, of course, you don't know the precisions of your procedures and you are fully aware that QED cannot be improved ad infinitum.

PHYS. The question of the accuracy of a procedure in physics is never of great importance. Somehow we always find that before we realize the mathematical limitations of a theory (or their absence !), we come across physical ones. Thus, for example, the mathematical limitations of QED for the γ, e system occur at $E \sim m_0 \exp(3\pi/2\alpha)$. This corresponds to $E \sim 10^{28}$MeV. In QED this energy, where the theory should lose its applicability, is usually denoted by Λ; since there are now many different Λ's, I shall call it Λ_{LP} in honour of Landau and Pomeranchuk [3].

MATH. But don't forget, the presence of Λ means that the theory cannot be precise just the same.

PHYS. That is true ! Logically I ought to say that the theory "should be changed" when we are in the Λ_{LP} region; and this would mean that if I were to set up an experiment at energy E, then I would have to make a correction of type $(E/\Lambda_{LP})^b$, where $b > 0$, or better still, $b \geq 2$. Obviously we don't have to worry about this.

MATH. I am amazed at your light-hearted attitude. How do you know that b is not equal to 10^{50} or, let's say, that it is not equal to 1 ?

PHYS. I know that it is not equal to 10^{50} from experiment; otherwise there wouldn't be any QED ! I know that b is not equal to1 from the model. I am talking about the method of making the theory finite, that is the regularization of Pauli-Villars [4]. I have calculated the correction in this regularization and I have obtained this sort of answer.

PHIL. If you have a method of making the theory finite, then why don't you come out and say that this is in fact the correct theory ?

PHYS. Because the Pauli-Villars regularization leads to nonsense when $E > \Lambda$, so that this is only a model after all. In fact, this is no longer a problem these days; physics has solved it !

MATH. How do you mean ?

PHYS. QED is now embodied in QFD and the latter is in GUT. GUT is asymptotically free at short distances, so there is no problem.

MATH. This is all very splendid but you know, of course, that GUT behaves badly in the Higgs sector.

PHYS. All right then. You make a cut off there and see whether the corrections are large at $E \to 30$ GeV. And you know perfectly well that the Higgs sector is a particularly phenomenological construct. I no longer want to regard the ultraviolet catastrophe as a current physical problem ...

PHIL. You have all mentioned on several occasions that the "series of QED" are asymptotic. It would seem that this is not as good as a "convergent series" but what is it ?

MATH. Well, you see, you can always expand a function near the origin in terms of its derivatives at 0 by writing down the series

$$f(x) = \sum_{n=1}^{\infty} \frac{1}{n!} f^n(0) x^n,$$

which sometimes converges and sometimes diverges. But every series is a code of some programme of computations, and if the series diverges, it is only the programme itself and the interpretation of the answer that changes. An asymptotic divergent series also enables one to say something about the function.

PHYS. Is an asymptotic series also a series in terms of the derivatives then ?

MATH. Certainly. The formal series $f(x) = \sum c_n x^n$ can be differentiated at the origin to obtain $c_n = \frac{1}{n!} f^n(0)$.

PHIL. I should like to know how and why a divergent series represents a function. I know about the geometric progression $\sum_{n=0}^{\infty} x^n$. I know how to work with it and I get

$$\sum_{1}^{N} x^n = \frac{1-x^{N+1}}{1-x} = S_N$$

and I understand that

$$S_N \to \frac{1}{1-x} = f(x) \text{ as } N \to \infty,$$

and by taking a sufficient number of terms I can for a given $|x| < 1$ obtain $f(x)$ to any desired precision. But what can one do if the series is divergent ?

MATH. An asymptotic series does not define a function uniquely. For a given asymptotic series there is an infinite set of functions corresponding to it.

PHIL. But after all, Phys has somehow calculated observable quantities by means of series that he somehow regards as asymptotic.

MATH. The Nth partial sum of an asymptotic series converges to $f(x)$ as $x \to 0$. Here the limit is taken not with respect to N as for your progression but with respect to x.

PHIL. I understand your definition but it's somehow not clear to me where actually such series are taken from and why, if the correspondence between a series and the function is not unique, its partial sum can approximate it just the same.

MATH. The simplest example I know of is unfortunately somewhat artificial. It is the series $\sum_{n=0}^{\infty} n!x^{n+1}$. It satisfies the differential equation $xf'(x)+x = f(x)$. You take this equation and you look for a solution in the form of a power series. Then you obtain the above series which is clearly divergent but which approximates the solution in the sense that I have indicated above. The situation with the differential equations of celestial mechanics is somewhat more complicated but similar.

PHYS. It's easy to understand why the series of QED are asymptotic. In the diagram of the nth order one more photon line can be inserted roughly in n ways; therefore the number of different diagrams increases as n! Therefore the series is similar to the series $\sum n!\alpha^n f_n(E)$.

MATH. Has this in fact been proved ?

PHYS. For certain simple Lagrangians.

PHIL. But why even so does a convergent series approximate the function and what is $f_n(E)$?

PHYS. I can take a concrete example and you will see that the difference between, say, S_1 and $f(x)$ tends to 0 as $x \to 0$. For example, this is simply done for the so-called $\Gamma(x)$ function defined by the integral $\int_0^{\infty} \exp(-u)u^{x-1}du$. I think that the situation here is rather like the QED situation where, the series of type $\exp(\int i\mathfrak{L}_{int}d^4x)$ also occur and $\mathfrak{L}_{int}$ contains the charge e as a factor. We are supposing that this series is asymptotic in e, that is, that in theory the first terms approximate the true answer to within an accuracy of $O(e^{2n+2})$. With regard to $f_n(E)$, I have written this in symbolic fashion; the answer of the nth order for the amplitude depends on the masses, energies, and momenta of the ingoing and outgoing particles, and in the general form, on all the Lorentz-invariant quantities that can be formed from all the four-dimensional momenta and all the masses, both external and internal.

MATH. I think there is nothing more to add here.

PHIL. But all the same I should like to obtain some sort of explanation why particles are quanta of fields. To understand why this is so, do I need to learn how to talk in this weird language of bundles and manifolds ? When I was younger, I read physics and mathematics over four semesters in the hope that this would be of some use to me.

PHYS. Splendid. I don't need much more. You know, of course, what a field is.

PHIL. Oh, certainly. For example, the field of the temperature in a body is given by a function $T(x,y, z, t)$.

PHYS. Yes, but in physics a field is often defined not by a single quantity, but several. Thus the Dirac field is defined by four complex quantities $\psi_1, \psi_2, \psi_3, \psi_4$.

PHIL. Aren't there any fields described by a single function ?

PHYS. One cannot give a simple answer with regard to real physics; so I suppose the answer is yes and no. But if we are talking about the formal structure of QFT, then it is possible to consider such a simple "field theory". This is the real "scalar field $\phi(\mathbf{r}, t)$ ".

PHIL. Splendid ! Can you "quantize" this field ?

PHYS. Let's try. But first of all we must clarify the classical theory of this field. Suppose that the evolution of a field ϕ with time is described by the equation

$$\frac{\partial^2\phi}{\partial t^2} - \frac{\partial^2\phi}{\partial x^2} - \frac{\partial^2\phi}{\partial y^2} - \frac{\partial^2\phi}{\partial z^2} + m^2\phi = 0$$

PHIL. I remember something of this sort from my physics course. I think that for this equation there's the abbreviated notation

$$\Box\phi + m^2\phi = 0.$$

PHYS. Even before the logical apparatus of QED, or even QM, emerged, Debye quantized proper modes of vibration as far back as 1910.

PHIL. What are proper modes ?

PHYS. We have a field ϕ dependent on $\mathbf{r}$ and t; but the fact that the space has three dimensions is for the moment unimportant. So suppose that we have a single spatial dimension and time. So what does our equation become ?

PHIL.

$$\frac{\partial^2\phi}{\partial t^2} - \frac{\partial^2\phi}{\partial x^2} + m^2\phi = 0 \tag{A}$$

I think I recognise this equation; it is the equation of the vibrating string.

PHYS. Not quite. The case $m = 0$ corresponds to the string. But there's no harm in your supposing that equation (A) describes a string fastened at the points 0 and L. Then the proper modes are the simple motions of the string for which ϕ changes according to the same law $\phi = A \sin \omega t$ at all points.

PHIL. But can these vibrations be properly represented other than as stationary waves of type 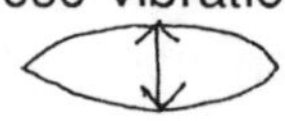 , with one antinode, or with two antinodes, and so on ?

PHYS. Yes. The general motion is a superposition of such motions described by the "Fourier series"

$$\phi(x,t) = \sum_{n=1}^{\infty} \sqrt{\frac{2}{L}}\, q_n(t) \sin\frac{\pi n x}{L}.$$

PHIL. I don't think there's any physics here. Any quantity $\phi(x, t)$ such that $\phi(0, t) = 0$, $\phi(L, t) = 0$ can be so represented.

PHYS. That's right, of course. But the physics consists in the fact that $q_n(t)$ satisfies the simple differential equation

$$\ddot{q}_n + \omega_n^2\, q_{\cdot n} = 0$$

where

$$\omega_n^2 = \sqrt{m^2 + \left(\frac{\pi n}{L}\right)^2}$$

PHIL. This equation is familiar to me. It's the equation of an oscillator. Its solution is, of course, $q = q_0 \sin(\omega t + \alpha)$, where q_0 and α are arbitrary.

PHYS. Yes, and such equations occur all the time in classical field theory when we expand in proper modes in two or three dimensions.

PHIL. But this is for vibrations in a cavity or vibrations of a membrane. Don't forget that in actual fact for elementary particle theory we need to consider fields in infinite space.

MATH. Theorists are always preoccupied with the ontological ghosts produced by the purely formal introduction of ideal objects. This is a very interesting phenomenon relating to the problem of formal and material existence.

PHYS. Well yes and no. Everything is both more complicated and simpler. Exp who has deserted us usually does his experiments in closed buildings, the walls of which perfectly absorb photons, hadrons and leptons (apart from the neutrino). So that the cavity approximation perhaps is an even better idealization.

MATH. But your beloved cavity corresponds to a reflection rather than absorption.

PHYS. Yes, but let's take a laboratory with reflecting walls. Yes, in the final event, perhaps the universe is finite. After I have finished my calculations for my laboratory lined with mirrors, I verify that the answer does not depend on its dimensions L if $L \gg \lambda$, where λ is the wave length of the particle.

MATH. (*to Phil.*) You have to get used to the amazing illogicality of the theorists and their reluctance to explain their models clearly. It helps them to work.

PHYS. With your permission, I will now write down the total energy of vibration of my string. It has the form $T+U$, where T is the kinetic energy and U is the potential energy; they are respectively equal to

$$\int_0^L \frac{1}{2}\left(\frac{\partial\phi}{\partial t}\right)^2 dx \quad \text{and} \quad \frac{1}{2}\int_0^L \left[\left(\frac{\partial\phi}{\partial t}\right)^2 dx + m^2\right] dx.$$

PHIL. The sum $T+U$ is preserved, by Newton's equality, but where do you get the coefficient 1/2 from ?

PHYS. This is a question of history or custom. The standardization of the constants is a matter of agreement. If we now go over to the q variables, then we obtain

$$E = T+U = \frac{1}{2}\sum(\dot{q}_n^2 + \omega_n^2 q^2)$$

PHIL. Yes I remember. I learned that the energy of the oscillator is always quantized and equal to $N\hbar\omega$, where N is an integer.

PHYS. Well, actually it is $\hbar\omega(N+\frac{1}{2})$, but the $\frac{1}{2}$ is not very important here. What this actually means is that according to quantum mechanics, the state of our string, of our "one-dimensional field", is described by prescribing all the N_n, the "excited states", for "all" the oscillators enumerated by the index n. The total energy is then given by

$$E(N_1, \ldots N_n, \ldots) = \sum_n \hbar\omega_n(N_n + \tfrac{1}{2}).$$

PHIL. So is this the basic result of QFT in accordance with this model ?

PHYS. That's right. But for the moment, we have not "deduced" this any more than Debye did, but "obtained it from experiment" as Planck did for the first time in 1900.

MATH. I recently read that Planck did not introduce discrete energy levels for the oscillator [5].

PHYS. Strictly speaking, he did not but I don't want to get involved in these historical niceties.

PHIL. Why are these "states of excitation" "particles" ? After all, the proper modes of vibration characterized by n are non-local. The field changes throughout the whole cavity. Aren't these some sort of symbolic oscillators rather than real ones ?

PHYS. If you go over from stationary waves to travelling waves and calculate the momentum of the field for the excitation $N_n = 1$, then you get $p^2 =$ $= \pm\sqrt{\omega^2 - m^2}$, depending on the direction of the wave, as according to the theory of relativity, this has to be the case for a particle with mass m and energy ω. (We are supposing once and for all that the units of length and time are chosen so that $h = 1$ and $c = 1$.) If we now introduce the interaction of the field with the absorber, then we find that when one quantum in the ith state is absorbed, the energy absorbed is ω_i and the momentum p_i.

MATH. But particles are absorbed and emitted "at a point", while in your description this is not the case.

PHYS. We can return to this when I have said something about how our system with energy

$$H = \frac{1}{2}\sum(\dot{q}_n^2 + \omega_n^2 q_n^2)$$

is formally described in quantum theory. But for the moment, it's worth remembering that as long as a quantum is characterized by a precise value of the energy-momentum (ω, p), it can in no way be associated with any point of space-time. This is the extreme case of the uncertainty principle.

PHIL. I feel with complete certainty that it's time to break off these explanations and have some tea.

Dialogue 9

PHIL. Last time, I understood that a field in a cavity can be represented as a system of non-interacting oscillators each of which, in accordance with quantum mechanics, has energy $E = \hbar\omega(N_n + 1/2)$. (*To Phys.*) But you then went on to appeal to experiment and the work of Planck. But surely there is in fact a consistent theory of "quantization" ?

MATH. Strictly speaking, there is an inconsistency in the very statement of the question in this form. Logically one should begin with quantum mechanics ...

PHYS. Indeed, we should begin with quantum field theory.

MATH. Yes, with quantum field theory and from that we can deduce classical field theory and from that classical mechanics.

PHYS. Well, even logically it is not as simple as all that. First of all, the relationship between quantum and classical mechanics is considerably more complicated than, say, that between physical (or wave) and geometric optics. Geometric optics can be "deduced" from physical optics as a method of describing a certain subclass of optical phenomena, valid under specified conditions; but in quantum mechanics the situation is more complicated. Expressions of a certain type such as: "the measurement of the quantity A for a system in state X will give the value A_i with probability w_i" are in general inexpressible in the language of quantum mechanics. Abstractly, we can enter the quantum world without having recourse to the famous "classical apparatus" of §7 of Landau and Lifshits. But this quantum world is quite unlike the real world that we have to deal with all the time and with which Exp sets up his experiments. Therefore in the canonical version of QM the "axiomatics" of the classical apparatus are contained in the theory as an independent part. It cannot be deduced from the remaining axioms [1]. In the second place, the very existence of stable matter (solid bodies or atoms) is a quantum effect. Therefore (*to Math*) a sequence of theories, where each is derived from the preceding one, does not exist.

MATH. Well, perhaps so. But even you must agree that if we know the Hamiltonian of a system we cannot uniquely define its quantum analogue; this is an ambiguous procedure and in that sense I am right. There are many quantum Hamiltonians that give the same classical one. This means that we need to begin straight away with the quantum Hamiltonian and then obtain the classical one from it.

PHYS. Maybe you are right in the general case. But historically we went from the classical oscillator to the quantum one and it seems to me that in this actual case there are no such problems.

In fact, history and experiment acquainted us with the classical oscillator to begin with and then, again through experiment, we guessed the rules of quantum theory.

MATH. It's remarkable how in this connection theorists were always dealing with fictions; the harmonic oscillators which Helmholtz, Lorentz and Planck dealt with are certainly not present in atoms.

PHYS. But you know full well that when a system (an atom) is close to its equilibrium state and you disturb it with light, then the deviation from equilibrium can be expanded into a sum of all the other states and the system is always like a system of oscillators. The successes of the theory of dispersion of Helmholtz and Lorentz were certainly no accident. The quantum theory of dispersion reproduces them very well [2].

MATH. Yes, but here the material oscillators with given frequency disappear; the things that represent them in quantum theory are "the transition from the ground state to a given excitation". It is very striking that when physicists were studying "the interaction of light with matter" they were studying non-existent oscillators.

PHYS. But after all, Debye found existing oscillators just the same, the proper modes of vibration in a cavity [3}. And the properties of Lorentz's fictitious oscillators are very like those of real transitions in dispersion theory. Lorentz was even able to construct a theory of the normal Zeeman effect by working with an electron oscillator.

MATH. Yes and this meant that the theorists from Poincaré to Landé and Goudsmit had to agonize about the anomalous Zeeman effect [4]. So in essence, the agreement of the Lorentz theory with experiment was accidental !

PHYS. Well, yes and no. Literally, his expression for this splitting of the spectral lines was correct to within a numerical factor !

PHIL. I think we had better come back to the Hamiltonian

$$H=\frac{1}{2}(\dot{q}^2+\omega^2 q^2)$$

What does its "quantization" look like ?

PHYS. I want to begin with yet another "fiction"; even the oscillator has an infinite number of levels 1, ...N, ... , but I want to describe to you quantum mechanics with a finite number of "levels" k. For such a system, the "state" is described simply as a k-dimensional complex column vector **c** with complex components $(c_1,c_2,...,c_k)^T$. An "observable" A is an "operator" which in this representation is simply a Hermitian $k\times k$ matrix A, that is, a matrix (a_{ij}) such that $a_{ij} = a^*_{ji}$ We say that the vectors **c** form a "Hilbert space of states of the system". (*To Phil*): I think you are familiar with the idea of a vector space ?

PHIL. Yes of course. But the appearance of these complex numbers bothers me; after all, in an experiment the measurements always give real numbers.

MATH. A state in quantum mechanics is not something that can be directly "observed" or "measured". The relationship is more complex here.

PHYS. Yes, although in fact there is no mystery here. We need to bring in another concept, that of the scalar product $<\mathbf{c}'\,|\,\mathbf{c}>$. Its definition is as follows:

$$\sum_{i=1}^{k} c_i'^* c_i = <\mathbf{c}'\,|\,\mathbf{c}>. \tag{9.1}$$

where the asterisk denotes the complex conjugate. Mathematicians denote it by a bar. You will notice that by contrast with the scalar product $\mathbf{a}\cdot\mathbf{b} = \mathbf{b}\cdot\overline{\mathbf{a}}$, it is not symmetric !

$$<\mathbf{c}\,|\,\mathbf{c}'> = <\mathbf{c}'\,|\,\mathbf{c}>^* \tag{9.2}$$

We now introduce the eigenvectors of the matrix A:

$$\sum A_{rl}\, c_l^{(a)} = A^{(a)} c_r^{(a)} \tag{9.3}$$

where $A^{(a)}$ and the index (a) enumerates the eigenvectors of the "operator" A. Our friend can prove that for any Hermitian A there are exactly k eigenvectors which are orthogonal in the sense of (9.1) and that all the A^a are real numbers. These vectors can be normalized so that $<c^a\,|\,c^a> = 1$. Any possible state of the system is described by some normalized **c**. The basic axiom then says: the measurement of a quantity A for the system in state **c** always gives one of the values A^a with probability

$$|<\mathbf{c}^a\,|\,\mathbf{c}>|^2 = w_a \tag{9.4}$$

PHIL. Is that all the axioms there are ?

PHYS. Well, strictly speaking, we also need to say that amongst the "observables" there is a distinguished one, namely, the energy H. This too is a $k \times k$ matrix. In the Schrödinger representation, the vector **c** varies in time according to the law

$$\mathbf{c}(t)> = e^{-iHt} \,|\, \mathbf{c}(0)> \tag{9.5}$$

PHIL. Why "in the Schrödinger representation" ?

PHYS. Because in the Schrödinger representation the matrices of the observables do not depend on time. Instead we can use the "Heisenberg representation". The states **c>** do not depend on time in this case and all the observables vary according to the law

$$A(t) = e^{iHt} A(0) e^{-iHt} \tag{9.6}$$

It is obvious that all the physical statements will be the same both in the Schrödinger and Heisenberg pictures.

PHIL. But how do we recognize what are observables and what are states in the system ?

PHYS. Generally speaking we guess this by relying on experiment.

PHIL. But we've got to start somewhere.

PHYS. That's right. What Heisenberg started with was the oscillator.

PHIL. But he had an infinite number of levels.

PHYS. You think of the matrices of observables as being infinite as well:

$$A = \begin{bmatrix} a_{11} & a_{12} & \cdots\cdots \\ a_{21} & a_{22} & \cdots\cdots \\ \cdot & \cdot & \\ \cdot & \cdot & \end{bmatrix} \tag{9.7}$$

but being like finite matrices in all other respects. In various places you will get convergent series or integrals instead of finite sums.

PHIL. That's fine. But what are the observables ? In the Hamiltonian we have q and $\dot{q}$. Now we have matrices, even infinite ones. How can we recognize them ?

PHYS. Well, I was taught as follows. You take the "canonical momentum p conjugate to q" and for the oscillator we obtain

$$p = \frac{\partial H}{\partial q} = \omega^2 \dot{q} \ .$$

You postulate the "commutation relation"

$$[p, q] = -i\hbar \tag{9.8}$$

You then express H in terms of p and q, obtaining

$$H = \frac{1}{2}(p^2 + \omega^2 q^2) \tag{9.9}$$

We can then define algebraically the matrices a, p and H from (9.8) and (9.9). Furthermore, we can make H diagonal, obtaining

$$H = \begin{bmatrix} H_{11} & & & & \\ & H_{22} & & 0 & \\ & & \ddots & & \\ & & & H_{NN} & \\ & 0 & & & \ddots \end{bmatrix}, \quad H_{NN} = \hbar\omega(N + \tfrac{1}{2}) \tag{9.10}$$

MATH. Strictly speaking, we have to state in advance that we are working in the basis where H is diagonal. Then ...

PHIL. I can imagine that the algebra will lead to this, but where does (9.8) come from ? More generally, where do these strange postulates about complex vectors and matrices come from ?

PHYS. Well, you see, theoretical physics has completely given up the hopes of the contemporaries of Plato, that everything is based on rational arguments. We apply some principles simply because they allow us to describe and predict experimental data. Relation (9.8) had to be adopted because for the oscillators of an electromagnetic field, it led to the formula $E_{N_i} = \hbar\omega_i(N_i + 1/2)$, which in turn led to the Planck distribution for the energy density of black-body radiation $\rho(\omega)$.

PHIL. I'm prepared to agree to all this, but there is no way back from the formula for $\rho(\omega)$ to (9.8).

PHYS. That's true. We have to introduce into the game lots more empirical and theoretical material. 25 years of strenuous research passed before Born and Jordan were able to guess at the relation (9.8).

MATH. I always thought it was Heisenberg who did this.

PHYS. That's true in fact. Heisenberg, in his famous 1925 article which marked the beginning proper of the creation of a consistent quantum mechanics, did not have in mind field oscillators, but again material oscillators. He dealt with a "toy" example; he tried to calculate the quantum anharmonic oscillator

$$H = \frac{1}{2}(\dot{q}^2 + \omega^2 q^2) + \lambda q^3$$

He guessed that q had to be regarded as a two-dimensional "array" and he guessed the rule of multiplication needed for calculating the powers. This was not sufficient to enable him to find q and Heisenberg wrote a relation for q and $\dot{q}$, which was equivalent to (9.8) [5].

In Born and Jordan's subsequent work it was realized that q had to be regarded as a matrix and that Heisenberg's multiplication rule was matrix multiplication, while Heisenberg's relation was equivalent to the formula (9.8) for the commutator. In this work, Born and Jordan also explained (in matrix form) the operator formalism of quantum mechanics for a system with one degree of freedom [6]. In essence, the description of quantum mechanics that I have been talking to you about is very close to the description in this article and in the subsequent article of Born, Heisenberg and Jordan. In this article vector states were also introduced. A very clear exposition of quantum mechanics from this point of view can be found in the Fermi lectures [7].

PHIL. Fine. So if I have properly understood this lesson, the state of an oscillator is described by an infinite-dimensional vector $\mathbf{c}_N$, where $|\mathbf{c}_N|^2$ is "the probability that the oscillator will be in the Nth excited state". In the general case the state is a superposition of all the excitations.

PHYS. That's right. So now if we consider "a field in a cavity", then this is a collection of an infinite set of oscillators indexed by numbers i and with frequencies ω_i. To define the state you need to specify the excitation number for all the oscillators; in other words, $\mathbf{c}$ is now a function of all the N_i, $\mathbf{c}(N_1,\ldots N_i,\ldots)$ and the energy of the given state is $\sum \omega_i N_i$ [8]. Dirac gave a complete description of the electromagnetic field in this form in 1927 [9].

PHIL. But surely the electromagnetic field is not a scalar.

PHYS. There are no great problems here. Essentially nothing changes.

MATH. Nothing changes as long as nothing is happening on the whole ! Your field does not interact with anything and is always in the same given state. If you introduce an interaction, then straight away you will see that the electromagnetic field is not a sum of oscillators.

PHYS. Some technical difficulties do occur, but they are easily avoided.

MATH. It seems to me that we have a serious problem here.

PHYS. Are you saying that you have doubts about the validity of the usual quantization procedures ?

MATH. Not for the electromagnetic field. But I am sure that it's absurd to proceed in this way for the Yang-Mills field.

PHYS. I don't think so, but time will tell.

PHIL. In Math's article I read about the creation and absorption operators. How are they described in matrix language ?

PHYS. That's easy. For the oscillator, the matrix q_i has the form

$$\sqrt{\frac{1}{2\omega_i}}\begin{bmatrix} 0 & \sqrt{1} & 0 & 0 & \cdots \\ \sqrt{1} & 0 & \sqrt{2} & 0 & \cdots \\ 0 & \sqrt{2} & 0 & 0 & \cdots \\ 0 & 0 & \sqrt{3} & 0 & \cdots \\ \cdots & \cdots & \cdots & \cdots & \cdots \end{bmatrix}$$

The remarkable thing about this is that if we let this matrix operate on the vector state $|\,0,0,\ldots N_i \ldots 0>$ corresponding to the i th oscillator in the N_ith excited state, then we obtain the state

$$\sqrt{\frac{1}{2\omega_i}}\left(\sqrt{N_i+1}\;.\;|\,0,0,\ldots 0,\, N_i+1,0> + \sqrt{N}\;.\;|\,0,0,\ldots 0,(N_i-1),0,\ldots>\right)$$

In other words, when it acts on a state with occupation number N_i, the operator q_i converts it into a superposition of states with occupation numbers N_i+1, N_i-1. Next we introduce the matrices

$$a_i = \frac{1}{\sqrt{2\omega}}(\omega_i q_i + ip_i)$$

$$a_i^+ = \frac{i}{\sqrt{2\omega}}(\omega_i q_i - ip),$$

where p_i is the matrix of the momentum for the ith oscillator

$$p_i = \omega^2 \dot{q}_i,\ \dot{q} = i(Hq - qH);$$

they will then have the form

$$a = \begin{bmatrix} 0 & \sqrt{1} & 0 & 0 & \cdots \\ 0 & 0 & \sqrt{2} & 0 & \cdots \\ 0 & 0 & 0 & \sqrt{3} & \cdots \\ \cdots & \cdots & \cdots & \cdots & \cdots \end{bmatrix},\quad a^+ = \begin{bmatrix} 0 & 0 & 0 & \cdots \\ \sqrt{1} & 0 & 0 & \cdots \\ 0 & \sqrt{2} & 0 & \cdots \\ \cdots & \cdots & \cdots & \cdots \end{bmatrix}$$

And these are the creation and annihilation operators. They have the property that (for a single oscillator)

$$a\,|\,N> = \sqrt{N}\;|\,N-1>$$

$$a^+\,|\,N> = \sqrt{N+1}\;|\,N+1>.$$

Upon acting on the unexcited state of the oscillator we obtain $a^+ | 0\rangle = = | 1\rangle$. As agreed, the state $| 1\rangle$ is the "field quantum". And that is all the secrets that there are. The Hamiltonian H of the field is expressed in terms of the operators a_i, a_i^+, where i is the "number of the oscillator", and has the form

$$\sum \tfrac{1}{2}\omega_i(a_i^+ a_i + a_i a_i^+).$$

When it acts on the state $| N_1, N_2, \ldots\rangle$, it gives the result

$$H | N_1, .., N_i, \ldots\rangle = \sum_i \omega_i(N_i + \tfrac{1}{2})$$

which corresponds to our intuitive ideas on the expected energy of such a state.

PHIL. But surely in your theory with Hamiltonian H "nothing is happening" in fact. Any state with energy E simply acquires the phase factor $\exp(-iHt)$ after a time t, and if we consider a superposition of states with different N_i, some sort of evolution occurs, but in general, "nothing is happening"; that is to say, the relative probabilities of states with different N_i do not change, while in fact the photons in the real world are emitted and absorbed.

PHYS. Well, this is quite simple. You know that in QED the source of photons is an antenna, an "external current". We add to H a term which is the interaction with the external current at the point x:

$$H_{int} = \phi(x)\mathfrak{J}(x)$$

Then H_{int} will cause a transition with absorption or emission of quanta. In fact Dirac in his article that I have already mentioned, introduced the "interaction" with the atomic electron rather than the "external current". Here the "transition" consists in an electron going over from one state to another and emitting a quantum of light; but in actual fact, this too is the completely developed version. In a consistent QED, the current $\mathfrak{J}(x)$ is also an operator containing the operators of creation and absorption of electrons and positrons.

MATH. I got the impression from the article by A. that the "canonical quantization" was inapplicable for electrons and positrons. Can you explain what is going on here ?

PHYS. First of all we need to say how we arrive at the field describing electrons. Its historical origins are quite different from those of the electromagnetic field. To put it briefly, for the electromagnetic field it was the "wave aspect" that was first recognized and then "by quantizing", the corpuscular aspect was obtained; in other words, by quantizing the electromagnetic field one obtained "light quanta" or photons.

With the electron it was all the other way round. To begin with, the electron was "a particle" and its "corpuscular aspect" was studied. Then de Broglie already recognizing the dual character of the photon associated the mysterious "de Broglie wave" with the electron. Schrödinger tried to consider the field ψ for an electron in its purely "wave aspect" and obtained the correct "non-relativistic wave equation" for it. Jordan and Wigner then showed that by quantizing the field ψ one could obtain a description suitable for a system of electrons described by numbers N_i by analogy with photons.

At the time when they were writing their article, the Pauli principle was already well known. There can only be in each given quantum state just one electron, therefore Jordan and Wigner realized that $N_i = 0,1$ for electrons. They therefore constructed a technique that would give this. They introduced matrices $b_i = \begin{pmatrix} 0 & 1 \\ 0 & 0 \end{pmatrix}$, $b^+ = \begin{pmatrix} 0 & 0 \\ 1 & 0 \end{pmatrix}$, where the state $\begin{pmatrix} 1 \\ 0 \end{pmatrix}$ is the vacuum of the oscillator and the state $\begin{pmatrix} 0 \\ 1 \end{pmatrix}$ is the "state with one quantum of the field ψ", that is an electron. In fact they also used some further information about electrons and came to the conclusion that while for the case of photons, the operators a_i satisfy the relation

$$a_e a_k^+ - a_k^+ a_e = \delta_{ke},$$

for electrons it is necessary to introduce the "anticommutation" relation [10]

$$b_e^+ b_k + b_e b_k^+ = \delta_{ek}.$$

PHIL. And where are the positrons ?

PHYS. That's a long story. To begin with one has to go from the Schrödinger equation to the Dirac equation for the "relativistic wave function" ψ with four components ψ_α. The field ψ, like the field ϕ, is a sum of modes. We then have to make ψ an operator just as we did with ϕ. In this case, there occur terms containing operators of absorption of an electron a and operators of creation of a positron b^+. This was arrived at by a long tortuous route. The modern text books on quantum field theory, of course, begin straight away by postulating the necessary form for the second quantized operators ψ, A (where A is the electromagnetic field operator) as well as other operators. After the fields, the commutation relations and the form of interaction have been given, one can derive the rule for calculating the transition amplitudes. These were stated in their most compact form by Feynman; I am referring to the famous

"Feynman rules" and "Feynman diagrams" [11].

PHIL. I can understand that historically some system of postulates, field equations, quantization rules has been arrived at, but it all looks very strange. Can't it somehow be explained more rationally ?

PHYS. I don't think your wish can be realized. After all, what we have discussed is the deepest level of our understanding of nature. It's impossible to explain our postulates. We can play with them and strive to understand better the proper nature of their potential implications, as did Bohr and Rosenfeld in the thirties [12]. We can elicit observable consequences from them and compare these with experiment; but we cannot ask "why are they like this and not like something else ?". Some clarification can be given here. For example, Pauli gave a plausible argument that if you try to quantize a field ϕ or A_i ("integer spins") with anticommutators or a field ψ ("half-integer spin") with commutators, then this is no good (some sort of potential internal contradiction). If you like, this is a partial explanation of "why this and not something else" [13].

MATH. Born and the probabilistic interpretation of ψ has somehow slipped out of your history of the electron.

PHYS. For single-electron states we can go from the Schrödinger equation to the "corpuscular aspect" simply by means of the probabilistic interpretation of $|\psi|^2$. And this is what Born did [14].

PHIL. You are all the time using "modes" that, so to speak, fill out the entire volume. But after all, in an experiment for example, photons are diffused in space, "absorbed at a point" and so on.

PHYS. In fact I introduced "interaction at a point" when I mentioned H_{int} with $\mathfrak{J}(x)$. I think you should try to read Fermi's celebrated article on quantum field theory. The spatial aspects of QED are worked out in detail there [15].

PHIL. I recall that in his article Math spent a long time considering how to construct vectors in Hilbert space for "bosons" and "fermions"; but you somehow seem to have avoided this !

PHYS. Yes. I don't think this is necessary for a logical application of the apparatus of second quantization. Math simply "started" not from this end although it's possible to enter the house that we live in even from his side.

I think one can completely define the theory by giving the commutation relations and the Lagrangian. There is no need to do anything beyond this.

MATH. In any case we have to begin with a complete list of fields and their group properties, but the gauge fields as we now understand them are not functions of points.

PHYS. Quite so. But in the present paradigm, this is essentially a superficial point with regard to the grammar of the theory just as the number of planets is a superficial point with regard to the "*n* body problem" of the mechanics of gravitation.

MATH. I think you are blasphemous.

PHYS. The blasphemy is in the way that the experiment is set up. The Apostle Thomas also blasphemed, or almost.

PHIL. I sometimes think that your paradigms are almost a religion for you!

Dialogue 10

MATH. (*To Phil*) Have you read the "recommended literature" ?

PHIL. I have looked at all the articles that Phys talked about: Fermi's article from "Review of Mod. Phys.", the article by Bohr and Rosenfeld, §7 of Landau's book, and those pages of the Pauli survey and Fermi's "Lecture Notes" that Phys mentioned last time. But all the same, I found it difficult to follow all the calculations and reasoning. Isn't it now possible, after such a long time, to set all this out somehow in a compact fashion and all in one place so that there would be a book on quantum field theory where the emphasis would be on the physical content rather than on the computational aspects ? I phoned Phys and asked him which book on quantum field theory was the best and he said Bjorken and Drell, but he added that although there's a lot of clearly written material on canonical quantization (in Vol.2) and the Feynman diagrams (in Vol.1), on the whole, ideas about the physical aspects of field theory, as discussed by Fermi and Born and Rosenfeld, are missing. Why is this ? After all, it is a necessary part of the theory.

MATH. Well, you see, when theorists create their theories, they discuss their "content" or "sense" or "meaning" among themselves or even independently but after a certain period all this goes off into some sort of "collective unconsciousness". Students study correct "theoretical behaviour"; that is the art of solving problems, in the way that mathematics was once taught in Babylon and Egypt and now in the engineering faculties. In mathematics, which takes as its model the Greek method of developing science, the problem of "meaning" is solved in a unified fashion: one considers systems of axioms and their set-theoretic models.

But even here, if you go into details, some pitfalls crop up. Cantor with his theory of the infinite and Hilbert with his confidence that infinity is after all just a symbol with defined rules of application, brought us to this precipice, while Gödel began to plot its boundaries on the map. We have insidiously got used to the idea that we are walking on a knife edge, and we have stopped worrying about it.

Out of habit we reproach the physicists for the inconsistency of their theory if they suggest that we should confine ourselves to certain distances in calculations. In a logical sense, the contradiction "at any distance" rules out the possibility of non-contradictory development.

There is a formal theorem that says that in any theory in which both statements "A" and "not A" are true, it is possible to prove any statement. However, the young in fact learn to steer clear of contradictions; otherwise none of them would ever become theoretical physicists.

PHYS. Apparently these ideas have well-defined archetypes. You have repeated H. Poincaré's train of thought. At the 1911 Solvay conference he said that all the participants acknowledged, apparently, that quantum ideas could not be combined with classical mechanics without contradictions; but this is just what the physicists had been doing; they attempted to get results by combining classical mechanics with some quantum notions. But once we admit a contradiction, then we can prove any statement [1]. In fact, the theorists gathered at Brussels in 1911 did not prove any theorems but tried to guess the quantum formulae by trying to find those occasions when classical physics worked, if only partially, and the quantum conditions were sufficiently simple. For example, they considered those occasions when "quasi-classics" (in modern terminology) were applicable. Subsequently, when the consistent non-relativistic quantum mechanics emerged, their actions obtained justification and their successes came to be understood.

PHIL. I find it difficult to understand how in such situations there can be any sort of confidence. I mean, the situation may turn out to be the same as in the old quantum theory; in other words, you have some algorithm that works for some chance reason but in actual fact, the things that Fermi, Bohr and Rosenfeld discussed, are all a lot of artefacts and the reality appears to be something quite different.

PHYS. How to solve the question of whether a theory still contains admixtures of artefacts or whether, so to speak, it "describes actual reality" as given by experiment, is a question which, although amenable to rational argument, is a difficult one. Here, errors of judgement are entirely possible. Say, as we now think, that the vacuum close to the nucleus is directly probed by a μ-meson. There's no artefact here to be sure [2]. The theory, by the way, predicts that starting with certain Z, this polarization leads to a change in Z which occurs with a jump. You introduce a nucleus with charge Z and the polarized vacuum screens it to Z-2.

PHIL. Why Z-2 ?

PHYS. That's a long story. The author of the article that I'm speaking about recently wrote a relatively understandable account of the question. You can read all about it [3].

PHIL. But I still haven't received an answer to the question why Bjorken and Drell don't contain in their book these matters that Bohr, Rosenfeld etc. were talking about.

MATH. You have to reconcile yourself to the fact that theorists one way or another deal with contradictory theories, but somehow avoid the contradictions. They are not capable of being discussed within the scheme of formal logic [4]. But they can nevertheless be discussed in the language of thought experiments. Thought experiments help them to understand the meaning of their statements better than real experiments. They always play an important role in the creation and formulation of theories. Look at Galileo the creator of mechanics. He is involved with thought experiments all the time. The role of Mach's thought experiments is well known, as are those of Einstein's in his creation of the general theory of relativity. In fact, they play an important role even in the creation of quantum theory, only they are hidden in the literature. For example, the adiabatic quantizations of the old quantum mechanics first appeared in the thought experiment discussed by Einstein and Lorentz at the first Solvay congress.

On the whole, I think that the thought experiment is roughly the unit of the "right-hemispherical mental process" corresponding to the deductive inference of the "left-hemispherical mental process". No doubt you have heard about the discoveries of Sperry in his experiments with the splitting of the brain. The properly functioning complete brain constructs images by bypassing mechanistic logic and verifies with logic the correspondence between the images and reality, so that reason does not fall alseep. Unfortunately, this does not get into the monographs and textbooks and it is clear why. The success of a theory is measured by its ability to describe actual experiments rather than thought experiments. At some point the books set out a computational scheme of the theory (say, QED) and compare the computations with experiment. Later on they stop doing this. It's not possible to find out from a textbook of theoretical mechanics even the accuracy with which, say, the motion of the planets is described. The fact is that when "everybody knows" that mechanics is true, then "nobody is interested" any more in the precision to which it is true.

PHYS. Well, I wouldn't say that. After all, when it was realized that celestial mechanics did not describe the precession of the perihelion of Mercury, everybody was very worried about it and it was even written about in the encyclopaedias [6] ! Everybody knew about this.

MATH. Show me a book on mechanics where they say something about the precession of Mercury and, more generally, about the accuracy with which mechanics enables one to calculate the motion even of the large planets where there is no relativistic precession.

PHYS. Unfortunately I don't know of any such book. "Ars longa, vita brevis". The art of writing textbooks progresses very slowly.

MATH. The quotation that you mentioned ends with the words "we cannot make a judgement on anything" [7], and yet you insist that you know that QED describes the facts and nothing but the facts !

PHYS. What is one to do ? We do not know, but we have to act just the same. In order to act successfully, we have to hang on to those assumptions that seem the most plausible. Of course, mistakes are always possible.

PHIL. All the same, I would very much like to see a book on quantum field theory where all the thought experiments which enable one to understand what the theory really is, are discussed. And apart from this, I should like to read there to what accuracy the calculations of the various quantities have been carried out. The actual methods of calculation are of no great interest to me. Obviously if you want to calculate to 10^3 figures, then this is a complicated undertaking. As far as I am concerned, the simplest cases are enough.

MATH. Yes, this would be a useful book.

PHYS. Is it worth writing ? Science changes so rapidly.

PHIL. Has it always changed rapidly ?

PHYS. Maybe not. Starting with Newton, the first paradigm of theoretical physics gradually crystallised: the mechanics of central bodies. It held sway for a long time. If you take a look at Poincaré's St. Louis address of 1904 [8], you will see that he implicitly asserts that theoretical physics works within the framework of this paradigm, although he recognized that it may have to be abandoned. Many people were aware of the inadequacies of the paradigm. It seems to me that Mach when he wrote his "Mechanics", "Theory of Heat" and "Optics" (published posthumously), wanted to write not so much philosophical books or the history of physics, but rather a physics course in which the paradigm would be critically discussed and analysed. He reverted to history as a method of recalling those thought experiments and logical discourses that the books of that time were so silent about.

MATH. So what happened ? Was the experiment successful ?

PHYS. Yes and no.

PHIL. Why yes and why no ?

PHYS. It's more of a question of *in what way* yes and *in what way* no. Mach didn't feel too badly about the inadequacies of Newtonian mechanics and his critical analysis, as is well known, was helpful to Einstein.

MATH. It is amusing that Einstein's favourite idea that he borrowed from

"Mechanics" is the very one that was not implemented. There is no "Mach principle".

PHYS. Surely that's not so amusing. The main thing about Mach's principle was in fact the correct "observation" that forces of inertia and gravitational forces are very similar. By working on this "principle of equivalence", but this time under relativistic conditions, Einstein with the help of thought experiments found certain effects of the general theory of relativity and, in the main, saw that he needed to go over to Riemannian geometry. The fact that the "field of gravitation" in the theory is defined not only by the sources (the heavy bodies) but also by the boundary and initial conditions proved to be unimportant.

MATH. Well, all right. But where is the "not" area ?

PHYS. The "not" is practically everything else. Mach was not a theoretician. Properly speaking, the theoretical physics of that time did not exist as a fully formed discipline, but there was "mathematical physics"; and this is not quite the same thing, or we can even say it is quite a different thing. Mach was an experimentalist, even a good one, but in a certain sense here, "being determines consciousness". The physics of the 19th century, to the extent that it possessed a self-consciousness, was, in my view, naïvely phenomenological and descriptively empirical, and the vehicle of this consciousness was in fact Mach. He did not understand the role of the theoretical construction of models and perhaps sometimes he simply did not completely know the theoretical physics that was already in existence. As a result, he denied the existence of atoms and conducted a fight against atomic theory which in some sense could be called desperate. Thus he had a completely false picture of the whole of physical theory and of the situation of his time.

MATH. Well, all right. Mach did not manage to make a critical and historical analysis of the paradigmatic situation at the end of the century and you think that we no longer need him since the paradigm of quantum field theory will soon change.

PHYS. No. The paradigm will not change by itself. It is precisely now that, if not a complete history, then at any rate a critical analysis of our paradigm would be very opportune. In the last decade there has been a striking success in QFT; the paradigm has turned out to be considerably more effective than had been hoped, but sometimes it seems to me that this is a Pyrrhic victory, as it were, and that the paradigm has reached its limits and maybe even overstepped them.

(*The telephone rings out for a long time. Phys goes into another room, listens briefly, and then returns.*)

PHYS. Exp has just phoned from CERN. Rubbiya has told him that he has seen the decay of $W^{\pm}$.

MATH. Well then! The paradigm is still holding up.

CHAPTER 2

Structure of Elementary Particle Theory

Go, go, go, said the bird:
human kind cannot bear very much reality.
(T.S.Eliot, Burnt Norton)

§1. Principles Underlying the Exposition

The aim of this chapter is to provide the material for the clarification of the notion of "formal existence of elementary particles". For the time being, therefore, we shall adopt the following point of view. The notion of elementary particle is meaningful only within the framework of the specified theoretical system (paradigm) being used for the description of nature.

The present paradigm is quantum field theory (QFT). Within its framework, elementary particles are quanta of fields which, for experimental or theoretical reasons we deem to be fundamental. The mathematical formalism of the theory includes a choice of the Lagrangian that is invariant with respect to gauge symmetries and is renormalizable; this, in principle, provides for the calculation of the fundamental quantities such as cross sections, spectra, decay widths, etc.

Every description of the universe offered by physics is approximate and phenomenological. However, with each penetration into the next level of "elementariness", there is always associated the hope that our knowledge will be of a more profound character, rather than having merely increased in extent. The laws of the next level are presented as being more fundamental then those of the previous level. When we give the theory a mathematical content, it often turns out that the passage to the theory at the new level involves a complete change in the fundamental mathematical structures used in its description. The essence of the special theory of relativity is not that it presents a systematic method of calculating small relativistic corrections to the classical laws of motion, but that it introduces the Poincaré group as the fundamental group of the space-time symmetries of physics. The main principles of quantum theory, which is the description of states as vectors in infinite-dimensional Hilbert space and the representation of measurable observables acting on them by Hermitian operators, does not in general have any roots in the preceding paradigm.

The very architecture of the hierarchy of approximate descriptions, once realized in such ideas as Bohr's "correspondence principle", becomes a part of the modern theory . In the empirical region it can be compared with the hierarchy of the characteristic scales of the processes under consideration. The notion of elementariness in its application to observable phenomena becomes relative: that which is represented as elementary in a given energy scale displays an internal structure under a finer resolution. The theoretical model of "elementarity" can turn out to be more stable. The same algebra of Fock space describes the elementary excitations in systems of different kinds and the photon, as an elementary particle, can at some level be regarded as equivalent to a phonon, that is, a collective excitation of vibratory degrees of freedom of a crystal lattice, since it is described by the same mathematics.

The opposition of formal and real existence arises in the interpretation of a theory, the syntax of which requires the introduction of "ideal elements" (and all deep theories are of this sort), and this is not the same as the "syntax of events", which is how it was represented in the light of the preceding paradigm.

In accordance with the aim of this chapter, QFT is described below not as a working instrument of the theorist but as an object of study. It is assumed that the physical principles known at present will turn out to have a broader domain of applicability than the specific theories. We have therefore attempted to explain the system of fundamental concepts in their modern mathematical form and to formulate the postulates on their relation to the reality. Naturally, much is sacrificed as a result; in the first instance, a working apparatus for computations.

Our account is synchronous in the main; in this chapter we only recall certain changes of ideas relating to the last two or three decades. We have picked out those items that are currently of value. One could have written a fascinating "history of the Lagrangian", tracing how the theoretical content of this concept, which arose long before QFT, has changed. But here the only thing of importance to us is the modern semantic field of the concept of Lagrangian. This refers also to the structures of space-time, symmetries and so on.

In conclusion we repeat: our theme here is elementary particles in the light of the present-day theoretical paradigm. The complementary point of view according to which the notion of the elementary particle is invariant with respect to a change of paradigm must be argued separately. This is an aspect of the idea of real existence.

§2. Elementary Particles and Interactions: Classifications

§2.1. The Classification and the Tables

In this section we discuss the systematization of the elementary particles and interactions presented by several tables and diagrams and commentaries on them. Classification plays a special role in the natural sciences, since it is the boundary region between the theory and the empirics and the region of their active interaction. Each table and each diagram is at the same time a code of empirical data and a code of the theoretical structure via which these data are interpreted.

In dealing here with the modern state of the theory of elementary particles we are more interested in the second aspect than in the first. The tables are therefore constructed so that their decoding enables one to denote certain essential codes of theoretical import. Thus, out of all the empirical data of hadrons, the two SU(3)-multiplets are selected; included in the examples of diagrams of fundamental interactions is the vertex of one of the hypothetical models of the grand unification etc. Adopting the terminology of the linguists, we can say that in the tabular and graphical material various levels of representation are deliberately mixed in an attempt to convey the intrinsic motion of the material. All the same, as far as the material itself is concerned, at this stage we approach it primarily from the descriptive aspect; the more systematic theory will be taken up later.

In the tables and their explanations, we have a fixed paradigm according to which the world is constructed from the elementary particles of matter, namely the quarks and leptons, that participate in the four types of interaction, strong, electro-magnetic, weak, and gravitational.

The material particles have spin 1/2 and are subject to Fermi-Dirac statistics. They are the quanta of the corresponding fermionic fields. They appear as point particles having no internal structure, although they do have internal degrees of freedom such as the colour of the quarks.

The interactions are carried by quanta of gauge fields that have spin 1 (gluons, photons, intermediate bosons) or 2 (gravitons) and therefore subject to Bose-Einstein statistics. These fields are analogous to the electromagnetic field as regards their action on the internal degrees of freedom of the particle. They bring about a rotation in the internal space just as the electromagnetic field rotates the quantum-mechanical phase. The difference is that the corresponding rotational groups for the weak and strong interactions are non-commutative. Moreover, the symmetry associated with these groups can be broken down, and one of the important mechanisms of the breakdown, the Higgs mechanism, predicts the existence of particles of a new type, namely the Higgs bosons with large mass.

The current picture of the classification and its theoretical bases is acknowledged by the physicists as being a transitional stage towards the theory of the next generation. This theory is called upon to unify all the interactions into a single scheme. The unified interaction can be governed by one symmetry group and one interaction constant. By way of illustration we systematically use the simplest SU(5)-model. In its literal form this model contradicts experiment because the prediction of SU(5) is in disagreement with experimental data on the determination of the lower limit for the life time of the proton; even so, the SU(5)-model can be regarded as a paradigm (in the linguistic sense) for the more complex models of the grand unification (GUT).

The ideas concerning the coupling constant, on which, in particular, the ordering of the interactions according to their strengths is based, is a somewhat complicated theoretical construct to which we shall return again. For the moment it suffices to say that in the scheme of quantum field theory, the coupling constants turn out to be variables; they depend on the distances or, in another language, on the transferred momenta characteristic for the process under consideration. If one allows oneself some daring exptrapolation, the various sizes of the coupling constant at energies attainable in modern accelerators merge at distances $\sim 10^{-29}$cm ($\sim 10^{15}$GeV), where the quarks and leptons "look the same" and the interactions become indistinguishable.

The postulate that the symmetry group is larger than that observed and that it is broken down by some mechanisms leads to the prediction of new particles, new interactions and new types of reaction. In particular, the symmetry between the quarks and the leptons predicts the possibility of proton decay and therefore the instability of matter.

In such a conceptual scheme, classification is in the first instance a recognition of the symmetry, latent or broken down, that is, the symmetry of laws that are latent or broken down in the phenomena.

The reader who has successfully managed part I will, it is hoped, be in no doubt that in the exact sciences, the motives from which the theory is applied by scientific society is radically different from the motives from which the theory is applied by the scientific society of mathematicians.

In the second instance the theory is a clear cut collection of postulates describing fragments of "reality" existing in the platonic world of ideas. For reasons that should apparently be regarded as inconceivable since Gödel's theorem, it is assumed that this set of postulates is irrefutable. The theory is a collection of consequences following from the list of postulates. In the ideal situation the theorems are proved and the proofs are accessible to anybody knowing the subject matter. The situation with the classification of finite simple groups shows that this aspect of the behaviour of the scientific society of mathematicians is also an idealization.

The theoretical physicist rarely worries about the presentation of the list of postulates of his theory for reasons discussed in the earlier pages.

The irrefutability of the theory is checked by mental tests; they in fact play an important role in the creation process of the theory itself. The theory in the narrow sense (excluding, say, Onsager's theory of phase transitions in the Ising model) should describe certain fragments of reality of the world of measurable phenomena; in the end, Phys must calculate quantities measurable by experiment. If in a given fragment of reality a sufficiently large number of results are obtained that agree with the results of observations, preferably with a maximally large number of digits and a minimum number of parameters of the theory, then the theory is accepted by the collective intuition of the society of physicists. Meanwhile, the theory may still be plagued by logical contradictions, as was the case in the "old" quantum theory of 1900-1925; it is suggested that these will subsequently be successfully eliminated. In fact Phys lives in such an unstable world and therefore numbers are so important for him.

In the world of modern theories, the champion is QED, regarded as the theory of photons and charged leptons. In principle we may take a fragment (A) describing the world e^+, e^-, γ, which with known and controllable accuracy can be isolated from the world (B): $(\mu^+, \mu^-, e^+, e^-, \gamma)$, of the world (C): $(\tau^\pm, \mu^\pm, e^\pm, \gamma)$ and finally, with very great precision from the world of hadrons.

As a result, in QED, (B) has three parameters m_e, m_μ and α. This last, as we think, is more "intelligible". The question "why" can be elicited from Chapter III. m_e, and m_μ are known to five figures and α to six.

There are standard sources of information on the numbers of the world of elementary particles. This is the "Review of Particle Properties", published by the international group "Particle Data Group", the number of participants of which is around twenty, and we have taken the information given below from that source. It follows from these (assuming that QED is good, although it may be self-contradictory), that the accuracy of the calculations in QED is of the order 10^{-5}. In reality, we can get rid of m_e from some of the numbers; the accuracy then becomes 10^{-6}. An example is known when simple-minded tricks, partly to do with words, bring the precision up to 10^{-9} (for the magnetic moment of the electron). The situation in QED is set out in more detail in a very clearly written booklet by M.A. Smondyrev "Quantum Electrodynamics and Experiment", *Znanie* , Moscow, 1984. The situation has not changed during the three years since its appearance; the reasons for this can be seen from Smondyrev's survey. The subsequent significant figures in physics are obtained with the same sort of difficulty as the final hundred metres on Everest.

The situation in the complete "standard theory" describing electrodynamic, weak and strong interactions is much worse. Roughly speaking, as far as hadrons are concerned, a struggle is going on to determine the first significant figure, and to achieve an indeterminacy of order two; for example, the mass of the gluon 0^{++}, varies in different calculations from 1 GeV to 2 GeV but is slowly shifting to the 2 GeV region. This does not contradict the fact that for the numbers involved in leptonic processes the accuracy is quite high and is determined by a knowledge of G_{weak} and m_μ for the calculation of τ_μ (see the end of Ch.III).

In "Particle Properties" the numbers are given in the form $m_\mu =$ $= 105.65932 \pm 0.00029$ or in the form $\alpha = 1/137.03604(11)$; in the latter case (11) denotes a ($\pm$) indeterminacy in the last two decimal places. The size of one Gaussian standard deviation is denoted by σ. The statistical procedure for collating all the data is complex. It is described in the book "The Fundamental Constants and Quantum Electrodynamics" by B.N. Taylor, W.H. Parker and D.N. Langenberg.

It should be recalled that the chances of an event occurring beyond the σ, 2σ, and 3σ limits are respectively 1/2.15, 1/21 and 1/370; this means that if an experiment is repeated under the same circumstances, the chances that a number will be beyond, say, 3σ, is the small quantity 1/370, but of course, this is by no means zero.

Since our tables are meant only to provide the reader with an acquaintance with the facts of particle physics, we sometimes round off figures so that the decimal places given can be regarded as trustworthy.

The reader should also be made aware that SU(4)-symmetries of hadrons in Nature do not occur, since the mass of a *c*-quark is much larger than the mass of the confinement Λ~(100-200) MeV. The three-dimensional weight diagrams provide a means, by virtue of their charm, for remembering easily which particles constructed from *u*, *d*, *s*, *c*-quarks are possible. The "etceteras" in the tables mean that the remaining decays, which have small relative probability, are not given.

TABLE 1

LEPTONS

	$Q=1$			$Q=0$ (neutrino)
	Mass (MeV)	Lifetime	Decay	Mass
Electron e	0.5	$>2.10^{22}$ years		ν_e 0-30 eV
Muon μ	105	2.10^{-6} sec	$e\nu\nu$	$\nu_\mu < 0.50$ MeV
Tau-Lepton τ	1784(3)	3.10^{-13} sec	$\mu\nu\nu$ 18(1)% $e\nu\nu$ 16(1)% $h\nu$ 48(2)% $3h+\nu$ 17(1)% (h hadron)	$\nu_\tau < 170$ MeV

HADRONS OCTET OF MESONS: $B = 0, J^{PC} = 0^{-+}$

	Mass (MeV)	Lifetime	Decay	Charge		Isospin	Hyper-charge	Strange-ness
				Q	I	I_3	Y	s
$\pi^{\pm}$	139.5	2.6×10^{-8}sec	$\mu^{\pm}\nu$	±1	1	±1	0	0
π^{0}	134.9	0.8×10^{-16}sec	$\gamma\gamma$ 99%	0	1	0	0	0
$K^{\pm}$	493.7	1.2×10^{-8}sec	$\mu^{+}\nu$ 64%	1	1/2	±1/2	±1	±1
			$\pi^{+}\pi^{0}$ 21%					
A further 15 decay channels are known								
K_S^0		0.9×10^{-10}sec	$\pi^{+}\pi^{-}$ 69%	0	1/2			
$K_0,\tilde{K}_0 = \pm$	497.7		$2\pi^{0}$ 31%					
K_L^0			$\pi^{+}\pi^{-}\gamma$ 0.2%					
		5×10^{-8}sec	$\pi^{\pm}e^{\mp}\nu$ 39%	0	1/2			
			$\pi^{\pm}\mu^{\mp}\nu$ 27%					
			$3\pi^{0}$ 21%					
			$\pi^{+}\pi^{-}\pi^{0}$ 12%					
Remaining channels ~1%								
η^{0}	549	0.7×10^{-18}sec	$\gamma\gamma$ 39%	0	0	0	0	
			$3\pi^{0}$ 32%					
and other channels			$\pi^{+}\pi^{-}\pi^{0}$ 24%					

OCTET OF BARYONS : B = 1, $J^P = 1/2^+$

	Mass (MeV)	Lifetime	Decay	Charge Q	Isospin I	I_3	Hyper charge Y	Strangeness S
p Proton	938.2	$> 10^{32}$years		1	1/2	1/2	1	0
n Neutron	939.5	~900sec	$pe\nu$ 100%	0	1/2	-1/2	1	0
Λ^o	1115.6	$2.6\text{x}10^{-10}$sec	$p\pi^-$ 64% $n\pi^o$ 36% and others	0	0	0	0	-1
Σ^+	1189	$0.8\text{x}10^{-10}$sec	$p\pi^o$ 52% $n\pi^+$ 48% and others	1	1	1	0	-1
Σ^o	1192	$5.8(1.3)\text{x}10^{-20}$sec	$\Lambda\gamma$ 99% Λe^+e^- 0.5%	0	1	0	0	-1
Σ^-	1197	$1.5\text{x}10^{-10}$sec	$n\pi^-$ 100% and others	-1	1	-1	0	-1
Ξ^o	1314.9	$2.9(0.1)\text{x}10^{-10}$sec	$\Lambda\pi^o$ 100% and others	0	1/2	-1/2	-1	-2
Ξ^-	1321.3	$1.6\text{x}10^{-10}$sec	$\Lambda\pi^-$ 100% and others	-1	1/2	-1/2	-1	-2

TABLE 2

A. Quanta of the fundamental fields of matter ($J = \frac{1}{2}$)

Leptons / Quarks ($J^P = \frac{1}{2}^+, B = \frac{1}{3}$)

	Leptons	Q	Quarks	Q	I	I_3	Y	S	(charm)
First generation	ν_e	0	u (upper)	2/3	1/2	1/2	1/3	0	0
	e	-1	d (lower)	-1/3	1/2	-1/2	1/3	0	0
Second generation	ν_μ	0	c (charmed)	2/3	0	0	-2/3	0	1
	μ	-1	s (strange)	-1/3	0	0	-2/3	-1	0
Third generation	ν_τ	0	τ (?)	2/3	0	0	0	0	0
	τ	-1	b (beautiful)	1/3	0	0	0	0	0

<u>Interaction carriers,</u>
<u>gauge field quanta</u>: | <u>Other particles</u>

γ (photon; electromagnetic) ($J = 1$) | H (Higgs bosons) ($J = 0$) (?)

g (gluon; strong) ($J = 1$) | X, Y (bosons) ($J = 1$) (?)

$W^{\pm}, Z^{o}$ (intermediate bosons; weak) ($J = 1$) | gravitino, gluino, photino,...

graviton (?) (gravitational) ($J = 2$) | ($J = \frac{3}{2}, \frac{1}{2}$) (? ?...)

B. <u>Quark composition of hadrons :</u>

Octet of baryons			Octet of mesons		
Σ^+	uud	$\frac{1}{\sqrt{2}}(udu - duu)$	π^+	$u\tilde{d}$	$-u\tilde{d}$
n	udd	$\frac{1}{\sqrt{2}}(uud - dud)$	π^o		$\frac{1}{\sqrt{2}}(u\tilde{u} - dd)$
Λ^o	usd	$\frac{1}{\sqrt{12}}(2uds - 2dus + sdu - dsu + usd - sud)$	π^-	$\tilde{u}d$	$\tilde{u}d$
Σ^+	uus	$\frac{1}{\sqrt{2}}(usu - suu)$	K^+	$u\tilde{s}$	$u\tilde{s}$
Σ^o	uds	$\frac{1}{2}(usd + dsu - sud - sdu)$	K^o	$d\tilde{s}$	$d\tilde{s}$
Σ^-	dds	$\frac{1}{\sqrt{2}}(dsd - sdd)$	$\tilde{K}^o$	$s\tilde{d}$	$-s\tilde{d}$
Ξ^o	uss	$\frac{1}{\sqrt{2}}(uss - sus)$	K^-	$s\tilde{u}$	$-s\tilde{u}$
Ξ^-	dss	$\frac{1}{\sqrt{2}}(dss - sds)$	η^o		$\frac{1}{\sqrt{6}}(u\tilde{u} + d\tilde{d})$

TABLE 3

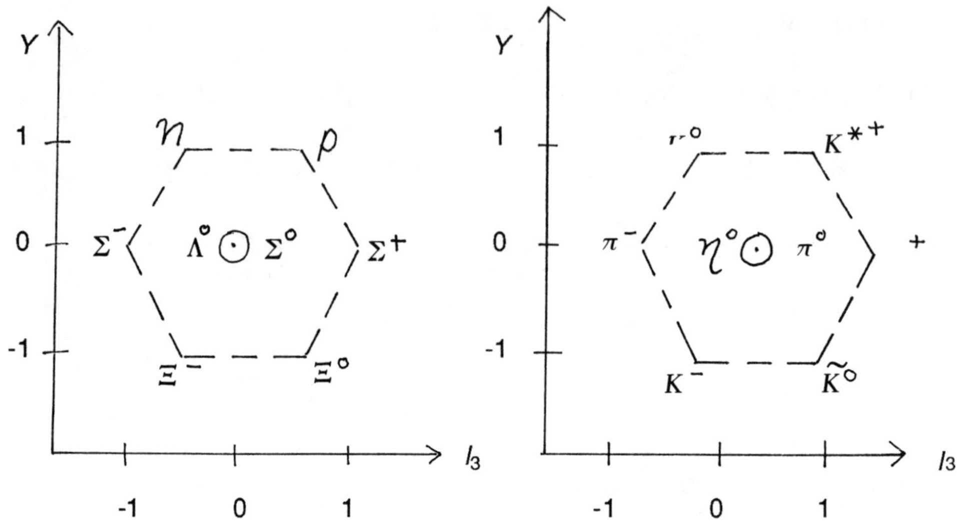

SU(3)-weight diagrams of the octets of baryons and mesons of Table 1

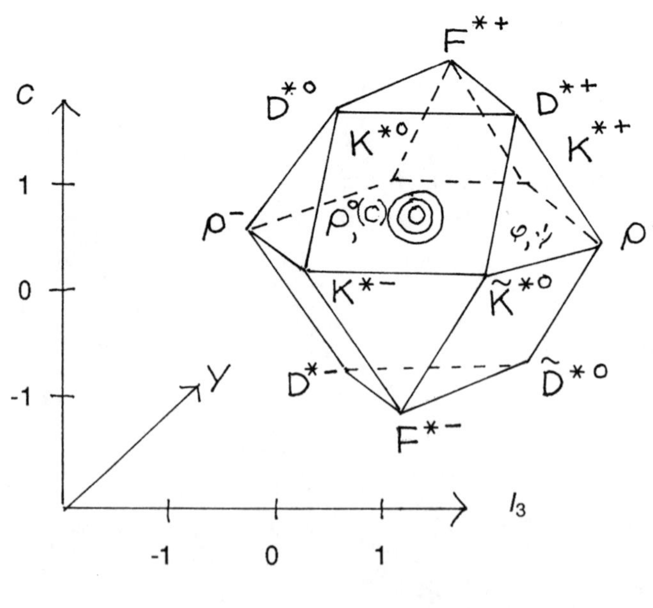

SU(4)-weight diagrams of the (15+1)-tuplet of vector mesons

TABLE 4

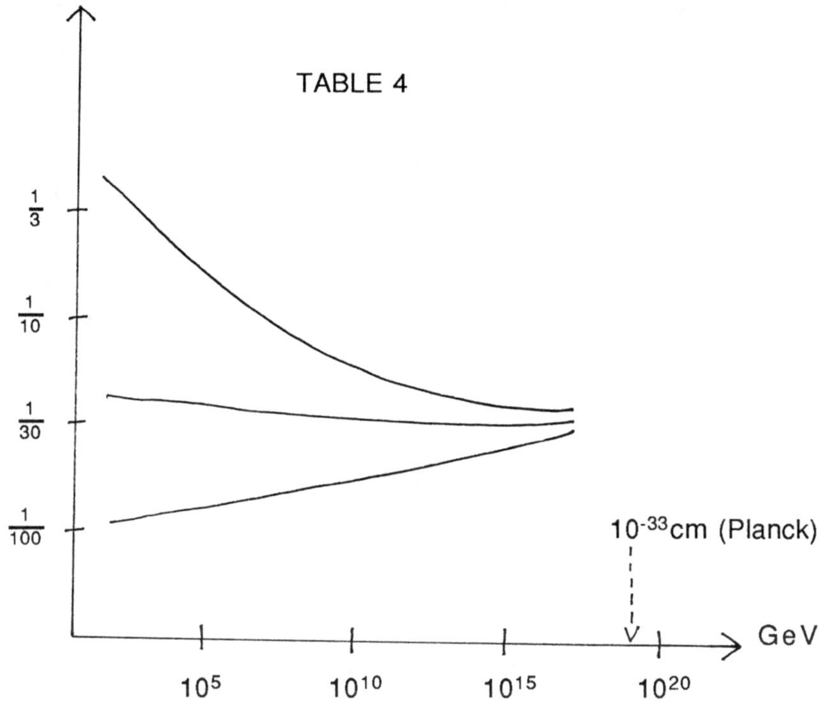

Running coupling constants

TABLE 5

Processes and interactions

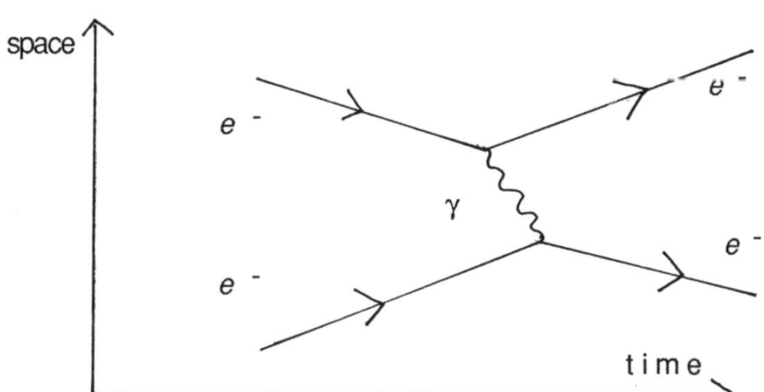

Electromagnetic:

Weak:

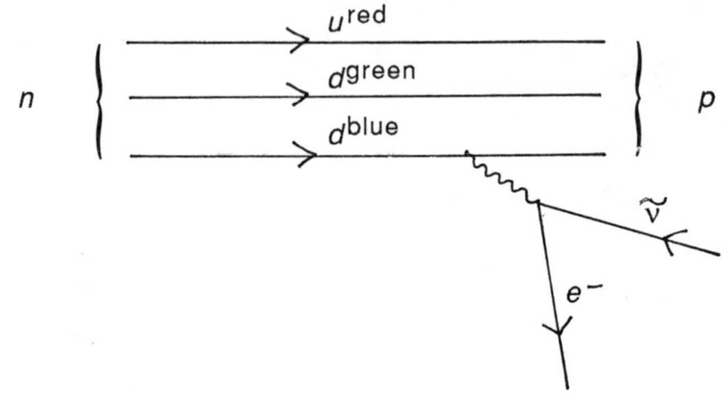

Proton decay in the SU(5)-theory

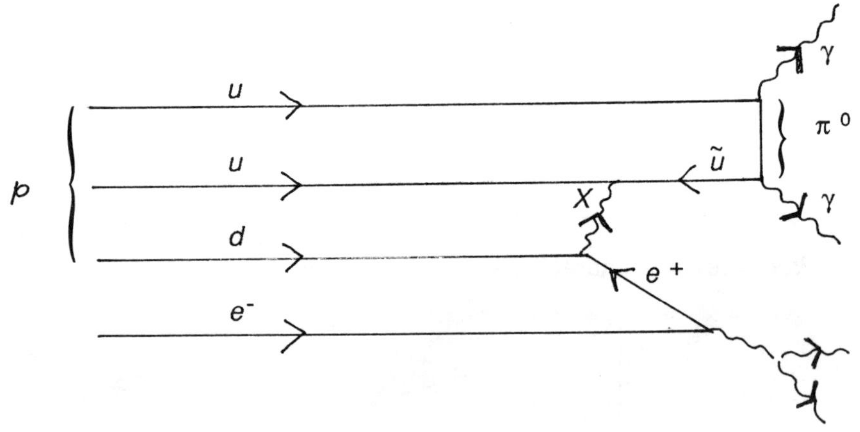

§2.2. Particles

Fundamental data on particles are given in Tables 1-3. In Table 1, we give for each particle its designation (sometimes a name) and its characteristics, which can be regarded as directly measurable: the mass, the average lifetime, the fundamental modes of decay and their probabilities (in percentages), and the charge. Besides this, we give the quantum numbers, such as J (spin), B (baryonic number) etc. Each cell of the Table does not of course directly reflect any empirical attribute and is in itself just a brief description of the result of the theoretical processing of a large number of observations including some that are by no means direct. If in this situation we allow ourselves to consider information concerning the lifetime and modes of decay of the original empirical matter, then it is only because more abstract concepts are a further object of analysis.

Table 1 does not contain all the experimentally observed particles: we have chosen from the large number of known hadrons only two multiplets of mesons and baryons; moreover, the antiparticles are not given for leptons and baryons; (we have put a tilde above them in our notation). Finally, we do not refer to particles that are carriers of the interaction; these are γ, $W^{\pm}$, Z^{0} and the graviton.

We now give a commentary on the matter contained in the Tables.

MASS

The mass of the particles is given in energy units (in accordance with the formula $E = mc^2$); the electron-volt is used as the unit of energy.

In the physics of elementary particles it is convenient to use a system of units in which the speed of light c and Planck's constant h are equal to unity. In this connection, the energy is measured in electron-volts, that is, the energy acquired by a particle with the charge of an electron on passing through a potential difference of 1 volt. The relationship between the rest energy of a particle $E = mc^2$ in the system $\hbar = c = 1$ has the simple form $E = m$, and the mass is measured in eV, or MeV or GeV. (1 MeV = 10^6 eV, 1 GeV = 10^9 eV.) The momentum p is also measured in eV; its value is the quantity pc measured in eV. The fundamental relations of quantum mechanics are the connection between the vibration frequency ν and the energy of a particle, and the momentum and its wavelength λ. In the usual units these relations take the form $E = \hbar\omega$ and $\lambdabar = \hbar/p$ or $\lambdabar = \hbar c/(pc)$, where $\lambdabar = \lambda/(2\pi)$ and $\omega = 2\pi/\nu$. In the system $\hbar = c = 1$, these relations take the form $\lambdabar = 1/p$ and $E = \omega$. Thus in these units there is just one dimensional quantity, namely the energy. Quantities having the dimensions of time and length have dimensions E^{-1} in this system. To express them in seconds or cm, one needs to multiply them by h and hc respectively, expressed in the units MeV sec and MeV cm. These take the numerical values 2×10^{-11} MeV cm and 0.7×10^{-21} MeV sec. For example, the so-called Compton wavelength of the electron $\lambda = h/(mc)$ is obtained in these units as $\lambda = 2\times10^{-11}$ MeV cm/(0.5 MeV) = 4×10^{-11} cm.

The old-fashioned crude classification according to mass is reflected in the etymology of the notation: lepton (light), meson (medium) and baryon (heavy). However, the tau lepton with its mass of 1784 MeV already shows that the mass can be large even for a weakly interacting particle. According to the latter criterion, tau also relates to leptons. The modern view is that leptons emerge as structureless particles while hadrons are connected systems of quarks. In the discovery of this fact, an important factor was that certain regularities had been discovered in the mass spectrum of hadrons; in particular, there are groups of hadrons, similar in mass, that are mass multiplets. The octet of baryons with masses from 938.28 to 1321.32 MeV is of this sort. It can be supposed, by analogy with atomic spectroscopy, that this multiplet yields a description in kinematic terms: as eight degenerate states of a single system that are split up by a small perturbation. The technical name of the corresponding mathematical scheme is $SU(3)_f$-symmetry; "3" refers to the three quarks u, d, s in Table 2, which within the framework of this approximate symmetry are considered to be equivalent: "f" is the flavour, which is the general name for types of quarks. If we do not distinguish between the quarks u, d, then they are combined into the multiplets (p,n) and (π^+, π^0, π^-): they correspond to the isospin symmetry $SU(2)_I$. All this will be explained in more detail in the next section. The masses of the quarks are theoretically a more complicated concept (due to their absence in a free state); one of the determinations gives $m_u \simeq 4$ MeV, $m_d \simeq 7$ MeV, $m_s \simeq 150$ MeV.

The neutrino mass column is a recognition of new trends. The mass ν_e is very small and until recently it was assumed that zero was the most plausible value; experiment gives upper estimates. Nowadays the conjecture is widely discussed that these masses are non-zero and, furthermore, there is a mass matrix giving the oscillations between neutrinos and antineutrinos of the same type as well as between neutrinos of different types.

LIFETIME

Apart from the electron, the proton, and possibly the neutrino, the remaining particles rapidly decay after their creation in accelerators or in cosmic rays. The rate of decay is usually well approximated by an exponential law $e^{-t/\tau}$; here τ is given in the tables as the average lifetime.

The characteristic nuclear time scale is 0.5×10^{-23} sec., the time in which the light traverses the Compton wavelength. In these units all the particles listed above are very long-lived. Among the hadrons that we have omitted are the considerably less stable ones (ω, ρ, ...); these were formerly called resonances. It is now clear that the entire systematization of hadrons and their characteristics is the classification of the excited states of quark systems, and it should be reduced to the more fundamental level of interactions between quarks, gluons, leptons etc.

Many of the decays can be put down to one of the types of interaction; thus, the muon decay $\mu^- \to e^- \bar{\nu}_e \nu_\mu$ occurs as a result of a weak interaction. In other cases several interactions participate in the decay; for example, the decay $\pi^0 \to 2\gamma$ is due to an electromagnetic and a strong interaction. The strength of the interaction is correlated to the speed of the decays; the weakest bringing about the slowest decays.

Of considerable interest are the neutral kaon-members of the meson octet. In reactions brought about by a strong interaction, such as $p+p \to K^0+\Sigma^+ +p$, they are created in the state K^0 and $\tilde{K}^0$ with quark state $d\tilde{s}$ and $\tilde{d}s$ respectively. Superposition (linear combinations) of these states K_S^0 and K_L^0 (short-lived and long-lived) possess definite masses and lifetimes in a vacuum. The need to consider superposition of particles and antiparticles, unique to kaons, is tied up with the weak processes that bring about the transitions $s \leftrightarrow d$. Finally, there are two further special superpositions K_1^0 and K_2^0, possessing a definite CP-parity (a combination of a charge conjugate C and a spatial P). In the K_1^0 state, a decay into $\pi^+\pi^-$ and $2\pi^0$ is possible with preservation of CP, and the decay $K_2^0 \to 2\pi$ is possible, the latter with non-preservation of CP.

QUANTUM NUMBERS AND CONSERVATION LAWS

A study of the list of possible decays and reactions in which there is more than one particle in the initial state, enables one to establish the kinematics of particles in the broad sense of the word, including the internal degrees of freedom. Here the postulate is applied according to which all processes not forbidden by the kinematics can occur; it is the dynamics that determine their probabilities.

The simplest expression of the kinematic laws of behaviour is the conservation laws. In their application to the spatial degrees of freedom they are as follows. A particle in the free state is characterized by its 4-momentum vector **k**. (In each inertial coordinate frame its spatial components correspond to the 3-momentum, and the time component to the energy.) In any reaction $a+b+...\rightarrow c+d+...$, the sums of the 4-momenta of the initial and final particles must be the same: $k_a+k_b+... = k_c+k_d+...$.

Since $k_a^2 = m_a^2$ (square of the mass), it is easy to see, for instance, that any particle can only decay into lighter ones. This, along with the law of conservation of electric charge $Q_a+Q_b+... = Q_c+Q_d+...$, explains the stability of the electron, the only ones lighter than it are the neutral particles, namely, the photon and the neutrino.

If we assign to all the quarks *u*, *d*, *s*, *c*, *b*, ... the baryonic charge $B = 1/3$, to the antiquarks $B = -1/3$ and to the leptons $B = 0$, and if B is calculated for hadrons on the assumption that B is an additive number, then it turns out that in all known reactions the baryonic charge is also conserved. By postulating this conservation law as universal and by appending it to the conservation laws for k and Q, we can explain the stability of the proton and, in the final reckoning, the universe as we observe it. An extrapolation upwards of the first steps of the ladder of broken symmetries $SU(2)_I \subset SU(3)_f \subset ...$ and the postulate of the symmetry group with respect to which the quarks and leptons are on an equal footing, lead to a breakdown of the law of conservation of baryonic charge and to the decay of the proton after 10^{31} years. In order to detect this effect, one would need to detect several proton decays in a year in several hundreds of tonnes of matter (10^{31} nucleons $\approx$ 16 tonnes).

Similar arguments have led in turn to the introduction of an additive number such as the strangeness S. For example, the absence of rapid electromagnetic decays of type $\Sigma^o \rightarrow p+\gamma$, $K^o \rightarrow 2\gamma$ etc., have been successfully explained by assigning the appropriate value of S to hadrons and by postulating that this value is conserved in electromagnetic and strong processes as well.

Nowadays these phenomenological quantum numbers, such as isospin, strangeness, charm, ... are directly interpreted in the quark model. We shall give their quantum-kinematic interpretation later.

CONCERNING TABLE 2

This table reflects the next level of theoretical generalization: with the exception of the photons and leptons, it lists the particles that are not in Table 1 and that have an essentially different experimental status. Thus, quarks (and gluons) experimentally appear as *partons* in the deeply inelastic scattering processes, say, of an electron by a proton. In this sort of process the momentum transfer and the common total energy of the secondary hadrons in the centre-of-mass frame must be » 1GeV (~the mass of the proton); it then turns out that the leptons interact not with the proton as a whole object, but rather with a point constituent of it carrying part of the 4-momentum of the proton; this constituent is called a parton. The history of the gradual identification of theoretical quarks (and gluons) with experimental partons is a very interesting illustration of the problem of formal/real existence.

The particles in Table 2 are sometimes called "fundamental", as opposed to "elementary".

The baryons and mesons represented in Table 1 are reproduced in Table 2 B along with their quark composition. These baryons are made up of the quarks u,d,s : the mesons from two and the baryons from three. The possibilities of such a naïve representation are limited; even without taking into account the space-variables, spin, and colour degrees of freedom, π^o (and η^o) has to be regarded as a superposition of states consisting of a quark and and an antiquark of two (or three) types. Finally, nucleonic states have an appreciable admixture of virtual quark-antiquark pairs (a "sea" of quarks, in contrast to the valency quarks that we have been talking about until now).

The quark c was discovered in the charmed mesons D^o, D^+, F^+; for us, they make their appearance in the last diagram of Table 3 with quark composition $c\tilde{u}$, $c\tilde{d}$, $c\tilde{s}$ and as a constituent of the famous particle J/ψ "with hidden charm" $c\tilde{c}$. The quark b together with $\tilde{b}$ makes up the upsilon particle, the meson Υ of mass approximately 9.4 GeV. The t -quark has yet to be discovered.

The immediate theoretical problem relating to quarks and gluons is to explain their confinement or, more precisely, their colour confinement, that is, the degrees of freedom responsible for the strong interactions. The nature of this force is such that it is effectively attenuated at small distances. This property of "asymptotic freedom" is understood theoretically.

The organisation of Table 2A itself, namely the division of leptons and quarks into generations and a comparison of the six known leptons with the six quarks, reflects certain symmetries of the weak, electromagnetic and strong interactions, partly already discovered and partly assumed. In the simplest theory of the "grand unification " based on $SU(5)$-symmetry, it is postulated that at energies greater than 10^{15} GeV the symmetry mixes up five and ten particles of each generation (say, $\gamma_{e,L}$; e_L^- ; $\tilde{d}_L$; $\tilde{d}_L$ is regarded as three particles on account of the three colours; the index L denotes "the left spin component").

Each postulated symmetry, if it is local, that is, if it can mix up the internal degrees of freedom at various points of space-time in different ways, automatically leads within the framework of Lagrangian quantum field theory to the prediction of gauge boson-particles, which carry the interaction responsible for this mixing. Of this type are the four particles $W^{\pm}$, Z^{0}, and γ in the standard unified model of electro-weak interactions or the gluons in quantum chromodynamics.

Under the "others" heading we have primarily included the undiscovered Higgs bosons *H*, which are used in the standard model for imparting a mass to the intermediate bosons. Possibly the introduction of *H* is a signal of the fact that we have stumbled across the next "effective" field which turns out to be a manifestation of the collective effects of some interaction at a more fundamental level. We then mention the *X* and *Y* bosons predicted by the $SU(5)$-unification and responsible for the possible non-preservation of the baryonic number. Finally we have listed the names of the particles with half-integer spins (gravitons, gluons) which are predicted by the supersymmetry and supergravitation theories. Meanwhile, the shadowy existence of these particles is not under threat of materialization.

CONCERNING TABLE 3

Table 3 consists of three diagrams which are very simply constructed. In the first two, points of the (I_3, Y)-plane are drawn corresponding to the particles of the octet of baryons and mesons respectively. In the third diagram the analogous figure is drawn corresponding to the 16-tuplet of mesons taking into account the three quantum numbers $(I_3, Y, C = \text{charm})$.

At the time when the contours of the spectroscopy of hadrons emerged based on approximate symmetry of the flavours, such diagrams provided a felicitous visualization of this symmetry in that they had the appearance of figures in the perfectly platonic sense.

If we are to be precise, however, these figures refer to another class of mathematically symmetric objects rather than the platonic polyhedra. Namely, they are diagrams of *roots* and *weights*. The simplest definition of root diagrams in purely symmetry terms is as follows : it is a finite system of vectors in Euclidean space (of any dimension) which is converted into itself under a mirror reflection with respect to any hyperplane orthogonal to one of the vectors (and satisfies a further integrality property).

Such systems were first discovered in the investigation of continuous symmetries, but were subsequently encountered by mathematicians in the most diverse, seemingly unrelated, problems. Their emergence in classification schemes of elementary particles can be interpreted as extra evidence in support of the fact that they represent "the archetypes of symmetry".

§2.3. Interactions

In the classical atomic theory, material particles are the centres of forces. The continuity of this idea in quantum theory is conveyed in the ideas on fundamental interactions. Interactions stipulate the existence of bound states, decays and reactions.

We briefly describe the role of interactions in the organization of the universe.

GENERAL CHARACTERISTICS

Nucleons, neutrons and protons are bound states of quarks. The binding is secured by the strong interaction between the colour charges of the quarks. This interaction is carried by gluons which themselves can have a colour charge (in contrast, say, to photons which bring about an electromagnetic interaction but do not carry an electric charge). The strong interaction rapidly increases with distance. This leads to the fact that particles with a colour charge, namely quarks and gluons, are always bound into systems of dimension of order 10^{-13} cm., with zero total colour charge, the so-called white states with which all observable hadrons are associated.

In turn, atomic nuclei are connected states of nucleons in which the binding forces are complex residual effects of the strong interaction.

The atoms are bound states of nucleons and electrons (leptons of the first generation); the binding forces are electromagnetic and the interaction is carried by photons. The molecules of ordinary matter are connected states of atoms. The binding forces are complex effects of electromagnetic interactions.

The primary interactions occurring in gauge theories are Coulomb-like, at least, at small distances; roughly speaking, the energy of the interaction of two charged particles is, to within numerical factors, e^2/r in the system $\hbar = c = 1$. This energy is written in the form α_{el}/r, where $\alpha_{el} = e^2/(hc)$, and e is the charge of the electron in ordinary units, the so-called fine structure constant. Similarly, the strong interaction at small distances has the form α_s/r, where α_s is the strong interaction constant. From a more advanced viewpoint, the quantities α_{el}, α_s themselves become functions of the distance between the particles.

The Coulomb potential describes forces acting at a distance. However, the electromagnetic charge is of two opposite signs, and in ordinary bodies of macroscopic dimensions, it is neutralized with high precision. In the form of electromagnetic waves (or, in corpuscular language, in the form of a stream of photons), electromagnetic interaction supplies us with the energy of the sun and almost all the attainable information concerning the universe beyond the limits of the earth.

The energy discharge in the sun occurs through nuclear reactions, the most important of which is considered to be the so-called *pp*-cycle in which four protons are converted into an He^4 nucleus with the emission of photons, positrons and a neutrino. Entering into this cycle are reactions with emission of neutrinos which are responsible for the weak interaction, for example, $pp \to de^+\gamma_e$ (where d is a deuteron). It was discovered in the course of elucidating the β-decay processes. The radius of the weak interaction is determined by the masses of the intermediate bosons m_W, m_Z, which are of the order 100 GeV. In the old theory, the dimensional Fermi coupling constant G was determined, for instance, from the average lifetime of the muon $\tau_\mu = 192\pi^2/(G^2 m_\mu^5)$. The calculation of G in terms of τ_μ gives the quantity $G \simeq 1 \times 10^{-5}/m_p^2$. In the Weinberg-Salam theory this quantity is expressed in terms of the electromagnetic constant α_{el} and m_W in the form $\pi\alpha_{el}/(\sqrt{2}m_W^2 \times \times \sin^2\theta_W)$, where $\sin^2\theta_W$ is a numerical parameter of the theory approximately equal to 0.2.

Finally, in systems of cosmic dimensions, the basic bonding forces are gravitational. In the current theory of elementary particles the gravitational interaction is not taken into account since, for example, for two electrons it is 10^{43} times smaller than the electromagnetic interaction. But the gravitational interaction is universal and induces just forces of attraction; when matter accumulates it becomes the dominant factor.

The gravitational interaction occupies a quite special position in the structure of the theory. The classical model of it, namely Einstein's general theory of relativity, is a space-time theory. Meanwhile, the remaining fields, quantum and classical, are fields in the space-time background. No consistent quantum account of gravitation is known; experimental quantum effects are far beyond the bounds of possibility for experimentalists.

RUNNING COUPLING CONSTANTS

As we have already said, the coupling constants in the theory are coefficients in the Lagrangians of the interaction. In the electro-weak model and the chromodynamics, Lagrangians include wave functions of gluons G, bosons W, a mixture of photons and Z^0-bosons of the combined model (B) and also quarks (q) and leptons (l). We write down these Lagrangians with constants g_i for a first acquaintance:

$$g_1\sqrt{\tfrac{3}{5}}B_\mu \sum_f \bar{f}\tfrac{Y}{2}\gamma^\mu f$$

(the summation is over all fermions *f*, quarks *q* and leptons *l*)

$$g_2 W_\mu \left(\sum_q \bar{q}_L \gamma^\mu \tfrac{\tau}{2} q_L + \sum_l \bar{l}_L \gamma^\mu \tfrac{\tau}{2} l_L \right)$$

(q_L, l_L denote the left spin components of the wave functions)

$$g_3 G_\mu \sum_q \bar{q} \gamma^\mu \tfrac{\lambda}{2} q$$

(Y, τ, λ are the generators of the symmetry group). The priming values of these coupling constants at attainable energies «100 GeV are as follows ($\alpha_i = g_i^2/4\pi$): $\alpha_1 \sim 1/67$, $\alpha_2 \sim 1/26$, $\alpha_3 \sim 1/5$. Denoting the value of the running constant at a given transfer of momentum M by $g_i(M)$, $\alpha_i(M)$, we obtain the theoretical formula

$$g_i^2(M) - g_i^2(\mu) = 2b_i \ln\frac{M}{\mu}$$

where M and μ, two transferred momenta, are much larger than the mass of all the particles, the contribution of which is taken into account, and $b_1 = -4$, $b_2 = 10/3$, $b_3 = 7$ (taking into account the three generations of fermions as in Table 2). In Table 4, we have shown the result of extrapolation of $g_i^2(M)$ into high energy regions: $\alpha_3(M)$ decreases, $\alpha_2(M)$ decreases, but slowly, while $\alpha_1(M)$ slowly increases; the point of convergence is in the region 10^{15} GeV. This is not all that far from the Planck scale where already gravitation has to be taken into account.

Apart from the actual presence of the point of convergence and the intrinsic attractiveness of the unified theory based on the group $SU(5)$, one also obtains from it a certain explanation of why three generations of fermions are "necessary" in the universe and the first one alone is insufficient. It is possible that the state of the world as we observe it, which is characterized by the sharp predominance of matter over antimatter has arisen as a result of the non-conservation of baryonic charge (and breakdown of CP-symmetry) at a very early stage (10^{-40} sec) after the Big Bang when the temperature was in the region of the unification energy.

FUNDAMENTAL PROCESSES AND DIAGRAMS

In the fifth and final table we give the forms of the Feynman diagrams.

Each diagram consists of lines and vertices where not fewer than three lines meet. The lines are divided into types corresponding to the fundamental particles of the theory (as well as to such objects as "ghosts", which we shall not discuss here). The vertices encode the elementary interaction processes postulated in the theory. From the naïve classification point of view adopted here, the list of all the possible vertices is a catalogue of which of the fundamental particles takes part in which acts of interaction and is therefore a catalogue of the interactions themselves as they are reflected in the given theory. Here "theory" is a specific Lagrangian, and the vertices to a first approximation correspond to the terms of the Lagrangian of degree higher than two (see §4).

For example, in the first three diagrams there are vertices corresponding to a) the emission and absorption of a photon by an electron; b) the emission of a gluon by a *u*-quark; this gluon changes the red colour of the *d*-antiquark absorbing the gluon into green; c) the emission of a W^--boson by a *d*-quark as a result of which the *d*-quark is converted into a *u*-quark of the same colour; d) the decay of a W^--boson with the emission of an electron and an antineutrino. Disregarding the details, these vertices correspond to the terms in the Lagrangian of the interaction in the previous subsection.

A diagram containing several lines and vertices can be used simply as a notation of a certain process. For example, in Table 5, the third diagram illustrates the basic features of the β-decay of a neutron. The neutron *udd* is converted into the proton *uud* as a consequence of the fact that one of the *d*-quarks is converted into a *u*-quark after emitting a *W*-boson which decays into an electron and an antineutrino. The last diagram illustrates the decay of matter into electromagnetic radiation in the *SU*(5)-theory: a *d*-quark of a proton emits an *X*-boson and is converted into a positron which annihilates with an electron of the matter; a *u*-quark of a photon absorbs an *X*-boson and is converted into a *u*-antiquark which forms with the remaining *u*-quark a π^0-meson which in turn decays into γ-quanta.

If, however, we look at which structures of the theory encode the Feynman diagrams, then the picture turns out to be much more complex. One of the aims of the theory is to calculate certain numbers, such as cross sections of processes, breadths of decays etc., in the same way as those mentioned in Table 1. These numbers are expressed in terms of amplitudes or matrix elements of the scattering matrix. The latter can be represented as a perturbation series, just as the diagrams correspond to the individual terms of them. The formulae for the running coupling constants are also calculated by means of Feynman diagrams.

Thus the diagrams are a visualization of a scheme of calculation that by no means claims to be fundamental. Furthermore, this calculation scheme, which is the road from the Lagrangian to the realistic numerical characterization of processes, with all its notable achievements, especially in quantum electrodynamics, comes up against a number of serious intrinsic problems.

Nevertheless, this scheme has proved to have had a decisive influence on the development of the theory in the last decades. The possibility of treating a Lagrangian by perturbative methods, the so-called renormalization of the Lagrangian, has come to be regarded as a principle guiding the choice of the theories. The calculation techniques connected with them have led to a number of intuitive images that physicists have come to use with great effect.

One of the most important images of this type is the notion of "virtual particles" which correspond to internal lines of the Feynman diagrams (similar to the lines of the photon in our first diagram). The structure of the amplitude corresponding to a single diagram forces us to interpret an internal line as a representation of a particle for which the relativistic formula $k^2 = m^2$, relating the energy, momentum and rest mass, is violated. However, since it is in existence for such a short time, the Heisenberg uncertainty relation does not allow this violation to be explicit. The logical extension of this concept is the representation of the physical vacuum as a medium in which creation and annihilation processes of virtual particles are occurring all the time; also the idea of a "dressed" particle, in contrast to a "naked" particle which is propagated by interacting with the vacuum and which is the best possible realistic approximation to the idea of a free particle.

All these images provide rich material for a discussion of the idea of formal existence.

§3. Quantum Kinematics

§3.1. The Kinematics

The subject of kinematics in classical mechanics is the study of the geometric properties of possible motions of bodies without reference to the forces and masses that determine the actual possible motions, the latter being the dynamics.[1] A product of the kinematical analysis of a system is a geometric model of it: the configuration and phase space of this system, also the coordinate functions on it, that is, the generalized coordinates and momenta.

In the broader sense, kinematics introduces those mathematical structures that are necessary for the description of the spaces of degrees of freedom of the systems under study. The mathematician considers the structures primarily from the geometric point of view, although the question here is no longer the geometry of three-dimensional Euclidean space where "the possible motions of bodies" occur; rather he is dealing with the symplectic geometry of phase space where there is a symmetry between the coordinates and the momenta, or the unitary geometry of the complex vector spaces of quantum mechanics. Finally, relativism enters into the kinematics of indivisible space-time (or momentum-energy) degrees of freedom with its geometry derived from the space-time geometry of Minkowski and its curved variants in the general theory of relativity.

Kinematics has taught us several lessons which were not apparent in the past. These are as follows :

a) *The presence of a small number of standard, fundamental internal structures in the spaces of degrees of freedom.*

In Hamiltonian mechanics these are the canonical coordinates and Poisson brackets. In quantum mechanics these are the superposition principle and the scalar products defining the amplitudes of the processes.

---

[1]Another way of saying this is that Euclidean geometry is the kinematics of solid bodies.

b) *The relative independence of certain fundamental degrees of freedom.*

Such structures as the mode of vibration and its quantum version, the quantum harmonic oscillator; polarizations, elementary vertices in particle theory, take part in the description of various systems in the same way as in natural language a small number of phonemes generate the various acts of speech. Analysis in terms of degrees of freedom leads to a more elementary level than analysis in terms of objects and processes.

In this connection, a theoretical tendency can be detected which construes elementary systems as systems that carry individual degrees of freedom "in pure form".

Probably the clearest example is still the notion of a mathematical point which kinematically is a pure carrier of the spatial degree of freedom. The Kant absolutization of space is a good example of the fossilization of a physical model in a philosophical dogma.

The visual perception transfers spatial relations into direct experience. Possibly one of the psychological predispositions of classical atomism was the naturalness of dividing up the universe into point-like elementary systems. The dual expansion into harmonic waves relates chiefly in perception to time and hearing, which provides less information about the universe; furthermore, the action of frequency analyzers is normally subconscious. The same thing applies to the facts discovered in the last decade, testifying to the fact that Fourier analysis takes place in the optical cortex, but its results are encoded before they are experienced.

c) *The importance of the kinematic symmetries.*

The kinematic symmetries act on the space of all possible states (motions). Each real process is in general much less symmetric, so that the kinematic symmetries are hidden.

Even so, they are responsible for such things as conservation laws (E.Noether). There is a close, and in elementary particle theory almost unique, relationship between the elementary degrees of freedom and the irreducible linear representations of the fundamental symmetry groups. Many quantum observables are infinitesimal generators of symmetry groups.

In conclusion we note that the kinematic description is essentially dualistic. One circle of ideas in it is connected with the idea of the internal state of an isolated system, rather than with an interacting system. A second group of ideas concerns that of interaction, namely the reaction of a system which also provides information about the internal state. The ideal of observation as a non-perturbing interaction arose probably in the bosom of observational astronomy. The rejection of this ideal has caused prolonged psychological struggles and epistemological arguments, but has proceeded irreversibly.

§3.2. The Superposition Principle

THE FUNDAMENTAL POSTULATES

The kinematic characteristics of an isolated quantum system or quantum degree of freedom are described by the following mathematical scheme.

a. The space of (pure) states is a set of rays of a complex vector space $\mathfrak{H}$. (We shall not touch upon the definition of the density matrix and mixed states that are necessary for an account of the classical incompleteness of information about the system.)

b. Corresponding to a joint system or a joint consideration of several degrees of freedom with state spaces $\mathfrak{H}^{(i)}$ is the tensor product $\bigotimes_i \mathfrak{H}^{(i)}$ or some subspace of it.

c. If a union of several systems with spaces $\mathfrak{H}^{(i)}$ has to be regarded as a single system, then we associate with it the direct sum $\bigoplus \mathfrak{H}^{(i)}$ of the original spaces.

d. Each state space $\mathfrak{H}$ is endowed with a Hermitian scalar product $<\chi|\psi>$. We set $|\psi|^2 = <\psi|\psi>$. The number $|<\chi|\psi>|^2(|\chi|\,|\psi|)^{-2}$ is then the probability that the system that was in state $|\psi>$ will now be in state $|\chi>$. The interpretation of this expression requires a postulate that there should exist a classical apparatus of a specific type capable of effecting the appropriate measurement.

We set out below the detailed mathematical and physical concepts installed into the postulates. We note the following important themes. The state of an elementary particle is described by several degrees of freedom (space-time, polarization, colour, ...) and the spaces corresponding to them are analysed separately. The inclusion of the quantum degrees of freedom describing the creation and annihilation of particles can be carried out in terms of the tensor algebra constructed over the single-particle state spaces as appearing in postulates b) and c). This "second quantization" mechanism is introduced in §3.4. Finally, we must not conceal the fact that in the relativistic QFT model, the existence of state spaces of a second-quantization field with the requisite properties has not been mathematically proved, and the calculations have been partly carried out within the framework of a reliably working formalism of algebra without a functional-analytic basis. Possibly this means that the standard ideology reflected in the postulates is unduly simple-minded: it reflects a part of the truth in its applications to fermions but requires considerable tightening for non-abelian gauge bosons. Meanwhile, we return to the details.

VECTOR SPACES AND SUPERPOSITIONS

A vector or linear space $\mathfrak{H}$ is a set whose elements can be added together and multiplied by complex numbers: if $\chi, \psi \in \mathfrak{H}$, then $a\chi + b\psi \in \mathfrak{H}$, where a, b are numbers. These operations must satisfy the usual rules of school algebra: $(a+b)\psi = a\psi + b\psi$, $a(b\psi) = (ab)\psi$ etc. The simplest example is the space of column (or row) vectors of length n, with coordinate-wise addition and multiplication. A second example is the space of complex functions of certain variables $x_1, \ldots, x_k$; extra conditions may be imposed on the functions, such as square integrability, satisfaction of some differential equation, and so on. The space of columns of length n is n-dimensional; function spaces are, as a rule, infinite-dimensional. The dimension of a space is the minimum number of basis vectors in $\mathfrak{H}$ such that any vector can be expressed as a linear combination of them.

A Hermitian scalar product on a linear space is a complex function of two arguments $<\chi \mid \psi>$ satisfying the conditions:

$$<\chi \mid \psi> = \overline{<\psi \mid \chi>}$$

(where the bar denotes complex conjugation):

$$<\chi \mid a\psi_1 + b\psi_2> = a <\chi \mid \psi_1> + b <\chi \mid \psi_2>; \quad |\chi|^2 = <\chi \mid \chi> > 0$$

(except for the case when $\chi = 0$). Here is a typical example:

$$<(x_1, \ldots, x_n) \mid (y_1, \ldots, y_n)> = \sum_{i=1}^{n} \bar{x}_i y_i .$$

Vectors ψ, χ for which $<\chi \mid \psi> = 0$ are said to be orthogonal; a vector such that $|\psi| = 1$ is said to be normalized. A space $\mathfrak{H}$ with Hermitian scalar product is called a Hilbert space if series of the form $\Sigma\psi_n$ with $\Sigma |\psi_n|^2 < \infty$, $<\psi_m \mid \psi_n> = 0$ converges to a vector in it. In the finite-dimensional case, this condition is automatically satisfied.

In the case when $\mathfrak{H}$ is a space of states of some quantum system, one can often indicate an apparatus or a process that will produce the system in one of its basis states (of the given basis characterizing this process). The inhomogeneous magnetic field in the Stern-Gerlach experiment splits up a beam of ions with spin 1/2 into two; this process gives a basis of the two-dimensional polarization space. The recording of the progress of the particle through the counter produces a state when its spatial coordinates are fixed. The idealized basis corresponding to this is described by the Dirac delta functions $\delta(x-x_0)$, where x_0 is the space-time mark of the counter; they can be regarded as a basis for the spatial degrees of freedom. In experiments in accelerators, the experimentalists prepare the particles with fixed values of momentum; this is another basis for the same space. (The mathematical niceties connected with the infinite-dimensionality are nowadays well understood and we omit them; the passage from one basis to another is a Fourier transformation. For a detailed discussion, see below.)

Given a vector $\psi \neq 0$, any vector $a\psi$ defines the same state (where $a \neq 0$). If we assume that ψ and $a\psi$ are normalized, the remaining arbitrariness $a = e^{i\phi}$ consists of the choice of the phase factor. Multiplication of the state vector by a phase factor does not change the state as long as we are not considering the interaction of systems, but is of the utmost importance in the description of interactions. It is often possible to represent the superposition of two states in the real normalized form $\chi = \cos\theta\cdot\psi_1 + \sin\theta\cdot\psi_2$, where θ is called the *angle of mixing*.

Some of the angles of mixing refer to some of the very important free (not given by the theory) parameters in the present-day models. For example, the Weinberg angle $\theta \approx 27^\circ$ enters into the description of the photons as the superposition of the two more fundamental fields A_3 and B: This superposition becomes a physically observable state of the electro-weak field after the Higgs fields break down the original symmetry. In the tables of the preceding chapter it was stated that the π°-meson is a neutral superposition of the quark and antiquark states $u\tilde{u}$ and $d\tilde{d}$ and that the η°-meson is a superposition of the states $u\tilde{u}$, $d\tilde{d}$ and $s\tilde{s}$ the mesons K_0 and $\tilde{K}_0$ are mutually orthogonal superpositions of the mesons K_L and K_S with different lifetimes.

Suppose now that the state space $\mathfrak{H}$ is represented as an orthogonal sum of subspaces $\oplus\mathfrak{H}_i$. It may turn out under specified conditions, or if some interactions are neglected, that not all the states $\psi \in \mathfrak{H}$ can occur as physically realizable states, but only as members of some of the sectors $\mathfrak{H}_i$. In such a case, the $\mathfrak{H}_i$ can be taken to refer to separate systems.

We now describe the most important "elementary degrees of freedom".

SPACE-TIME

The classical space of the states of a material point is the physical Euclidean space $\mathbf{R}^3$. In Schrödinger's quantum mechanics it is replaced by a space of complex-valued functions $\psi(x)$, $x \in \mathbf{R}^3$. Each such function can be regarded as a (continuous) superposition of Dirac delta functions: $\psi(x) = \int \psi(x')\delta(x-x')$. By supposing that $\delta(x)$ describes a quantum state corresponding to the image of a particle localized at the point O, we discover that the Schrödinger description is (almost) a corollary of the two postulates : a) superposition, and b) the existence of states that are precisely localized in the space ("correspondence principle").

We now consider an absolutely delocalized quantum state, that is, one that is not altered by a spatial translation. The corresponding ψ-function satisfies the functional equation $\psi(x+a) = e^{if(a)}\psi(x)$, where the phase $f(a)$ is easily seen to be a linear function on $\mathbf{R}^3$, that is, $f(a) = p \cdot a = p_k \cdot a_k$. Therefore $\psi_p(x) = e^{ipx}$ where the vector p, which is uniquely determined by the state ψ_p, is called the *momentum* of the particle in this quantum state. The scalar product $p \cdot x$ is measured in Planck units of action $\hbar$ and is therefore just a real number. The characteristic wavelength associated with a particle in a state ψ_p is $2\pi/p$, therefore in acts of interaction with such a particle the spatial structure of the target is resolvable at distances of order $1/p$ (only by p one means the transferred momentum, that is to say, the fraction of the momentum that actually participates in the interaction).

This same argument applied to space-time $\mathbf{R}^4$ leads to a distinguished class of states of plane-wave type $e^{-ik \cdot x}$ in relativistic kinematics, where on this occasion $k = (p,E)$ is the 4-momentum, and $x = (x_1, x_2, x_3, t)$; $k \cdot x = -p_i x_i + Et$, E is the energy part (Et being measured in the same $\hbar$ units). Under a change of the inertial coordinate frame, k changes but $k^2 = E^2 - p_i p_i$ (the square of the mass of the particle) does not.

The states of plane-wave type are frequently prepared by experimentalists, however, with regard to the fundamental particles, they succeed only with leptons. Quark plane-wave states can only be respectable approximations at small distances where there is asymptotic freedom; the so-called hadron beams in deeply inelastic processes provide evidence in favour of their reality. At distances of order of the dimensions of nucleons, where the quarks are bound in stable formations, it is possible to try to use spherical waves as basis vectors; these will be locked in some small volume like a bag. In a consistent theory, of course, things should become more interesting mathematically.

Thus, the general scheme of dealing with space-time degrees of freedom is as follows: $\mathfrak{H}_{sp}$ is a space of functions of space-time coordinates. Any such function is a field, so that the state of the system has the character of a field, even if the system consists of particles. In the future theory the gravitational interactions must be accounted for in such a way that in the classical limits the gravitational interactions must be described by Einstein's curved space-time. Such a general space-time does not possess symmetries enabling one to introduce plane (or spherical) waves, and we lose many of the usual facilities of the computational apparatus of perturbation theory. Nevertheless, the majority of the constructs of QFT in the background of the classical curved space-time can be carried out in a consistent fashion just the same.

POLARIZATIONS

The spin or polarization degree of freedom is characterized by a quantity that has the dimensions of angular momentum. It has to be regarded as an internal degree of freedom since the structureless fermions have spin 1/2, that is, internal angular momentum $\hbar/2$. It is, however, closely tied up with the space-time characteristics; in particular, the double covering of the symmetry group of space-time acts at the same time on the polarization space. None of the remaining "purely internal" degrees of freedom possess this property. We list the main characteristics of the spin state space for $J = 1/2$.

Associated with each point of the state space is the complex 4-component space of Dirac bispinors $\psi = \begin{pmatrix} u \\ v \end{pmatrix}$, where u, v are two-component spinors. The spinor ψ satisfies the Dirac linear equation of the form

$$i\,\partial_k \gamma^k \psi = m\psi$$

where γ^k is the four-dimensional matrix satisfying the conditions

$$\gamma^j\gamma^k + \gamma^k\gamma^j = g^{jk}, \quad g^{jk} = \text{diag}(1, -1, -1, -1).$$

Under a suitable choice of the form of the matrix γ^j, the spinor v vanishes identically in the rest frame of the particle, while corresponding to spinors u of the form $\begin{pmatrix} 1 \\ 0 \end{pmatrix}$ and $\begin{pmatrix} 0 \\ 1 \end{pmatrix}$ are two spin states, as must be the case for a particle with $J = 1/2$.

In the description of massless fermions, it is more convenient to use another representation of the matrices γ in which the spinors u and v are transformed independently under Lorentz transformations. Here the spinor matrix $\begin{pmatrix} u \\ 0 \end{pmatrix}$ (symbol 2_R) corresponds to the state of particles with spin along the direction of motion, while the spinors $\begin{pmatrix} 0 \\ v \end{pmatrix}$, or 2_L, correspond to spin opposite to the direction of motion. (These are the so-called right and left particles and spinors.)

The same Dirac equation also describes the antiparticles.

ISOSPIN

We can associate with a particle with isospin $I = 0, 1/2, 1, 3/2, \ldots$ a $(2I+1)$-dimensional internal space. In the original Heisenberg version, the space corresponding to $I = 1/2$ was introduced as a space of superpositions of a neutron and a proton and was subsequently applied to the mathematical statement of the notions of n and p being indistinguishable in strong interactions. In the modern paradigm this is the space of superpositions of a u-quark and a d-quark.

WEAK ISOSPIN

In the weak interactions described according to the Weinberg-Salam model, the degree of freedom called weak isospin makes its appearance. The states in the weak isospin space 2_W are represented in the lepton sector by doublets of left-polarized leptons $\begin{pmatrix}\nu_e\\ e\end{pmatrix}_L, \begin{pmatrix}\nu_\mu\\ \mu\end{pmatrix}_L, \begin{pmatrix}\nu_\tau\\ \tau\end{pmatrix}_L$. In the quark sector there arises the interesting phenomenon of mixed generations. It turns out that the quarks entering into the doublet 2_W and the physical quarks u, c, t and d, s, b are not the same. If we denote by the symbols $\begin{pmatrix}u'\\ d'\end{pmatrix}, \begin{pmatrix}c'\\ s'\end{pmatrix}, \begin{pmatrix}t'\\ b'\end{pmatrix}$ the three equivalent 2_W-representations, then the physical quarks, which are states with a given mass forming part of the hadrons, are obtained as superpositions that are independent for the upper and lower quarks. Thus, say, $d = \alpha d' + \beta s' + \gamma b'$ etc.

COLOUR

The strong interaction of quarks is brought about by an exceptionally important degree of freedom corresponding to the three-dimensional colour space 3_C for the particles u, b, e, s, ... and $\overline{3}_C$ for their antiparticles. The three basis vectors of this space are conventionally denoted by the triple of base vectors "red, yellow, blue" (and for $\overline{3}_C$ the corresponding "anticolours"). These degrees of freedom are associated with the carriers of a particular colour charge.

It is important that in 3_C there are no physically distinguished state vectors; in the language that will be explained later, this means that the colour symmetry is not broken down by anything. Thus, the names "red, yellow, blue" are conventional not only with regard to their standard semantics, but in the following deeper sense: they do not denote concrete vectors in 3_c in precisely the same way as "length, breadth and height" do not denote actual directions in physical space (although in the laboratory this symmetry is violated by the earth's gravitational field, which distinguishes "height").

DEGREES OF FREEDOM IN COMBINATION

According to postulate b) at the beginning of this section, in order to construct the state space of a system taking into account its various degrees of freedom, we have to form the tensor product of the spaces corresponding to these degrees of freedom.

We give a brief description of this mathematical construction.

Let $\mathfrak{H}_1, \ldots, \mathfrak{H}_m$ be vector spaces. Then their tensor product contains all possible elements of the form $\psi^{(1)}\otimes\ldots\otimes\psi^{(m)}$ (where $\psi^{(i)}\in \mathfrak{H}_i$), called decomposable, and all their superpositions (i.e. linear combinations). In a decomposable state the system has a completely determinate state in each degree of freedom: $\psi^{(i)}$ in $\mathfrak{H}_i$. If the degrees of freedom correspond to subsystems, then in decomposable states these subsystems preserve their individuality; the ith is in its state $\psi^{(i)}$. Any part of the universe lends itself to a classical analysis with a decompostion into component parts only to the extent that its quantum state is close to a decomposable one.

A tensor product of the vectors $\psi^{(1)}\otimes\ldots\otimes\psi^{(m)}$ is linear in each of its arguments and no further conditions are imposed on this operation. Therefore, for example, a basis in $\mathfrak{H}_1\otimes\ldots\otimes\mathfrak{H}_m$ can be constructed in the following way: we choose a basis in each of the spaces $\mathfrak{H}_i$ and in each such basis we choose a vector and then tensor multiply all these vectors together. The existence of such a decomposable tensor basis shows that the dimension of the tensor product is equal to the product of the dimensions of the component spaces. (In the infinite-dimensional case one has to worry about infinite series and convergence; we leave this out.) The scalar products on $\mathfrak{H}_1$ and $\mathfrak{H}_2$, say, define the scalar product on $\mathfrak{H}_1\otimes\mathfrak{H}_2$ according to the formula:

$$<\psi^{(1)}\otimes\psi^{(2)} \mid \chi^{(1)}\otimes\chi^{(2)}> = <\psi^{(1)} \mid \chi^{(1)}> <\psi^{(2)} \mid \chi^{(2)}>.$$

If $\mathfrak{H}_i$ is the space of row vectors $(\psi_k^{(i)})$, where k runs through the indices enumerating the basis vectors in $\mathfrak{H}_i$, then $\mathfrak{H}_1\otimes\ldots\otimes\mathfrak{H}_m$ in the tensor basis is the space of m-dimensional matrices $(\psi_{k_1\ldots k_m})$, where each k_i runs through its index set. In the Dirac notation the k_i can be quantum numbers and instead of $\psi_{k_1\ldots k_m}$ one writes $\mid k_1\ldots k_m>$ etc.

In the case when one considers the tensor product of m identical spaces, then there are two important subspaces of $\mathfrak{H}^{\otimes m} = \mathfrak{H} \otimes \ldots \otimes \mathfrak{H}$ which are defined by symmetry properties under a permutation of the indices.

The space $S^m\mathfrak{H}$, the mth symmetric power of $\mathfrak{H}$ is generated by the tensors of the form $\frac{1}{m!}\sum_\sigma \psi^{(\sigma(1))} \otimes \ldots \otimes \psi^{(\sigma(m))}$ where σ runs through all the permutations of the indices. Similarly, the space $\Lambda^m\mathfrak{H}$, the mth exterior (or antisymmetric or Grassmann) power of $\mathfrak{H}$ generated by the tensors of the form $\frac{1}{m!}\sum_\sigma \varepsilon_\sigma \psi^{(\sigma(1))} \otimes \ldots \otimes \psi^{(\sigma(m))}$, where $\varepsilon_\sigma = +1$ for even permutations and -1 for odd permutations.

For example, $\mathfrak{H} \otimes \mathfrak{H}$ splits up into a direct sum of two sectors: the symmetric $S^2\mathfrak{H}$ and the antisymmetric $\Lambda^2\mathfrak{H}$. When $m > 2$ there exist tensors with more complex symmetries with regard to their properties relative to permutation groups; however, these play a considerably less significant role than $S^m\mathfrak{H}$ and $\Lambda^m\mathfrak{H}$.

Tensors can be multiplied together: $(\psi_1 \otimes \ldots \otimes \psi_k) \otimes (\psi_{k+1} \otimes \ldots \otimes \psi_s) = \psi_1 \otimes \ldots \otimes \psi_s$. Thus we can introduce the tensor algebra of the space $\mathfrak{H}$: it is the direct sum of all the spaces $\mathfrak{H}^{\otimes m}$, where $m \geq 0$ and where $\mathfrak{H}^{\otimes 0}$ is simply the complex numbers.

The product of two symmetric tensors is not in general symmetric, but it can be symmetrized. One then obtains an operation of commutative and associative multiplication on the symmetric algebra $S(\mathfrak{H}) = \bigoplus_{m=0}^{\infty} S^m\mathfrak{H}$ of $\mathfrak{H}$. Similarly one can define the exterior or Grassmann algebra $\Lambda(\mathfrak{H}) = \bigoplus_{m=0}^{\infty} \Lambda^m\mathfrak{H}$: the exterior product of two antisymmetric tensors is the result of antisymmetrizing their tensor products.

In physics $S(\mathfrak{H})$ and $\Lambda(\mathfrak{H})$ are called the Fock spaces; later on we shall discuss their role in the formalism of the second quantization.

In the actual calculations, an essential role is played by the rules for calculating the various combinations of tensor products and symmetrization. Rather than give a systematic account here, we mention the simple example;

$$\Lambda^2(\mathfrak{H}_1 \otimes \mathfrak{H}_2) = \Lambda^2\mathfrak{H}_1 \otimes S^2\mathfrak{H}_2 \oplus S^2\mathfrak{H}_1 \otimes \Lambda^2\mathfrak{H}_2 .$$

In other words, a 2-tensor that is antisymmetric in the aggregate index (ik) can be uniquely decomposed into a sum of two tensors, the first of which is antisymmetric in i but symmetric in k, and the other way round for the second tensor.

We now turn to physical examples.

THE FUNDAMENTAL PARTICLES

The Dirac particles with spin 1/2 and without extra degrees of freedom are described by the space $\mathfrak{H}_{sp\text{-}t}\otimes(2_L \oplus 2_R)$ (where $\mathfrak{H}_{sp\text{-}t}$ is the space-time part of the wave functions). The neutrino is of this type; if a neutrino, say, ν_e, is massless, then in the interactions that we consider, only the left particles participate and we must confine ourselves to the sector $\mathfrak{H}_{sp\text{-}t}\otimes 2_L$. To take into account the electric charge in *a* of the units e^+, it is convenient to introduce the one-dimensional internal space 1^a_{em} with the following rules for tensor multiplication:

$$1^a_{em} \otimes 1^b_{em} = 1^{a+b}_{em} \quad \text{and} \quad \overline{1}^a_{em} = 1^{-a}_{em} \; .$$

The electron-positron single-particle space will then be $\mathfrak{H}_{sp\text{-}t}\otimes(2_L\otimes 1^{-1}_{em} + 2_R\otimes 1^{1}_{em})$. For the needs of the weak interaction, the lepton and its neutrino are combined in each generation into a left weak isodoublet with space $\mathfrak{H}_{sp\text{-}t}\otimes 2_L\otimes 2_W$. A quark of specified type, say *k*, corresponds to the space

$$\mathfrak{H}_{sp\text{-}t}\otimes(2_L\otimes 1^{2/3}_{em} + 2_R\otimes 1^{2/3}_{em})\otimes 3_C.$$

Decompositions of state spaces into sectors and tensor products similar to those that we have described, often turn up in another context, as a hierarchy of group representations and symmetry breakdowns. We shall return to this point of view in the next section.

UNIFICATION OF IDENTICAL PARTICLES

Associated with *m* identical systems in the state space $\mathfrak{H}$, is not the complete tensor product $\mathfrak{H}^{\otimes m}$, but the subspace selected by the symmetry conditions with respect to permutations.

If the particle is a boson (with integral spin), then its *m*-particle space is $S^m\mathfrak{H}$; while if it is a fermion (half-integer spin), then the space is $\Lambda^m\mathfrak{H}$. For example, in the non-relativistic description of the two-electron cloud in the helium atom we have the state space $\Lambda^2(2\otimes\mathfrak{H}_{sp})$, where 2 denotes the two-component Pauli spin space and $\mathfrak{H}_{sp}$ the space of Schrödinger wave functions. It decomposes into a sum of two sectors: the singlet with respect to the spin $\Lambda^2(2)\otimes S^2\mathfrak{H}_{sp}$ and the triplet with respect to the spin $S^2(2)\otimes\Lambda^2\mathfrak{H}_{sp}$. These two sectors are orthogonal and passage between them is kinematically inhibited, and this is reflected in the singularities of the helium spectrum.

After half a century since this classical argument, an analogous argument has become one of the most important pieces of evidence in favour of the existence of the highly mysterious space 3_C. Here is a simplified version of it. There exist baryonic states of three identical quarks with spin 3/2 and orbital angular momenta that give evidence of the symmetry of the state wave function with respect to permutations of the quark coordinates (an example is the particle $\Delta^{++} = uuu$). If we ascribe to u the space $\mathfrak{H}_{sp}\otimes 2$ and assume that the quarks are subject to Fermi statistics, then we find that the wave function Δ^{++} lies in the sector $S^3\mathfrak{H}_{sp}\otimes ?$ of the space $\Lambda^3(\mathfrak{H}_{sp}\otimes 2)$. It is not difficult to see that this sector, with wave function that is symmetric with respect to the coordinates, is equal to zero because of the two-dimensionality of the spin space. The introduction of 3_C saves the situation by enabling one to locate Δ^{++} in $S^3(\mathfrak{H}_{sp}\otimes 2)\otimes\Lambda^3 3_C$. It should be noted that the space $\Lambda^3\, 3_C$ is one-dimensional: the colour state corresponding to it is red$\wedge$blue$\wedge$yellow, that is "white". In the meson two-quark sector, $\overline{\text{red}}\otimes\text{red}+\overline{\text{blue}}\otimes\text{blue}+$ $+\overline{\text{yellow}}\otimes\text{yellow}$ is called the white colour vector. Note that this vector does not change even if we call (red, blue, yellow) some completely different orthogonal basis in 3_C. The subspace of $\overline{3}_C\otimes 3_C$ orthogonal to the white vector describes the eight colour degrees of freedom of the gluons: by convention we say that the gluon $\overline{\text{red}}\otimes$blue bonds together q^{red} and $q^{\overline{\text{blue}}}$ by forcing these quarks to exchange colour charges.

The kinematic law of the colourlessness of connected states of quarks ought to have a dynamical explanation in quantum field theory. The presence of three colours is supported by calculations according to perturbation theory where the three-colour property leads to the so-called cancellations of anomalies, as well as by the behaviour of the ratio of the cross sections of the reactions $\sigma(e^+e^-\to\text{hadrons})/\sigma(e^+e^-\to\mu^+\mu^-)$ in their dependence on the total energy at the centre of mass. The theoretical formula gives the following quantity for this ratio: (*number of colours*)$\times$(*the sum of the squares of the charges of quarks that are created free at a given energy*). The experimental graph shows plateaux corresponding to the three colours and values $R(u, d, s) = 2$; $R(u, d, s, c) = 10/3$ (sum of squares of the charges).

§3.3. Symmetries and Observables

GENERAL INFORMATION

Let $\mathfrak{H}$ be the space of states of a quantum system. To characterize each individual state in classical physics one postulates the possibility of measuring the value in each state of certain physical quantities such as energy, coordinate, momentum etc. In Hamiltonian mechanics, the mathematical model for such quantities, the *observables*, is provided by the differentiable functions in the phase space of the system. In quantum mechanics, a mathematical model of observables is provided by Hermitian linear operators on $\mathfrak{H}$, that is, linear maps of $\mathfrak{H}$ into itself with a real spectrum that are diagonalizable in an orthonormal basis of $\mathfrak{H}$. Then, while a classical observable assumes a determinate value at each state of the system, a quantum observable itself defines a set of states on which its values are determinate: these are the eigenstates and eigenvalues of the operator. A quantum observable A is associated with some device that measures this observable. Such a measurement then gives one of the eigenvalues of A for the corresponding observable.

Each observable f in Hamiltonian mechanics is not only one of the coordinate functions in phase space, but also a generator of a one-parameter group of canonical transformations, or phase flow. A motion with respect to time is the phase flow generated by the special function H, the Hamiltonian. The remaining canonical transformations are the kinematic symmetries of phase space; they preserve the form of the Hamilton equations of motion, but, in general, they alter the Hamiltonians and trajectories.

In exactly the same way, each quantum observable A is a generator of the one-parameter group e^{itA} of transformations of $\mathfrak{H}$ that are linear and preserve the transition amplitudes. Here the motion with respect to time is also defined by the Hamiltonian or energy operator.

It has proved to be fruitful to take the view that the symmetry group of a quantum system (or Lie algebra of its infinitesimal generators) is the primitive mathematical object. The state space $\mathfrak{H}$ is then constructed as a linear representation of this group, while the generators (or some functions of them) turn out to be the fundamental observables.

We now embark on a more detailed description of the mathematical and physical aspects of this scheme.

UNITARY GROUPS AND HERMITIAN OPERATORS

Let $\mathfrak{H}$ be an n-dimensional complex space with Hermitian scalar product serving as the space of the degrees of freedom of some quantum system. We consider all transformations of this space that are linear (compatible with the superposition principle) and preserve scalar products. These transformations form the unitary group $U(n)$. The groups $U(n)$ are the basic groups of the kinematic symmetries of quantum systems. The subgroup of unitary transformations with positive determinant (in fact equal to +1) is denoted by

$SU(n)$ (here S stands for "special"). In an orthonormal basis $U(n)$ consists of the $n \times n$ complex matrices V such that $V V^+ = 1$, where $V^+ = \overline{V}^T$, in other words, $V_{ik}^+ = \overline{V}_{ki}$.

The group $U(1)$ consists of those complex numbers whose modulus is 1, that is, the phase factors $e^{i\omega}$, where ω is a real number. Similarly, each element of $U(n)$ can be expressed in the form e^{iX} where X is a Hermitian operator. Each Hermitian operator X defines some orthonormal basis of $\mathfrak{H}$ in which the matrix of X is diagonal and real. For a fixed orthonormal basis, each Hermitian operator is represented by a matrix X such that $X^+ = X$. These operators do not form a group but a real Lie algebra $u(n)$, that is, a real linear space that is closed with respect to the bracket operation $\frac{1}{i}[X, Y] = \frac{1}{i}(XY - YX)$. A unitary matrix e^{iX} belongs to SU if and only if X has zero trace: $\operatorname{tr} X = X_{kk}$ (summed over k) $= 0$. The Hermitian matrices with zero trace form a Lie algebra $su(n)$. We now give the explicit form for the bases of $su(2)$ and $su(3)$ used in particle theory.

a. *The Dirac and Pauli Matrices*

A basis for $su(2)$ is formed by the Pauli matrices

$$\sigma_1 = \begin{pmatrix} 0 & 1 \\ 1 & 0 \end{pmatrix}, \quad \sigma_2 = \begin{pmatrix} 0 & -i \\ i & 0 \end{pmatrix}, \quad \sigma_3 = \begin{pmatrix} 1 & 0 \\ 0 & -1 \end{pmatrix},$$

with commutation relations

$$\frac{1}{\sqrt{-1}}\left[\frac{\sigma_i}{2}, \frac{\sigma_j}{2}\right] = \varepsilon_{ijk}\,\frac{\sigma_k}{2}; \quad \varepsilon_{123} = 1,$$

where ε_{ijk} is totally antisymmetric. If we consider a massive particle with spin 1/2 having magnetic moment in its rest frame, then in the two-dimensional space of its spin states, those states in which σ_k is diagonalized correspond to the stationary states of the particle in a magnetic field directed along the kth axis. In particular, the basis in which the matrix is written consists of states "with spin projection 1/2 and -1/2 along the x_3 (or z) axis". As we have already mentioned, this rule of interpretation relates the space-time degrees of freedom with polarizations.

If $\mathfrak{H} = 2_I$ is the isospin space of a particle with isospin 1/2 (p, n, Ξ^0, Ξ^-; u, d, see the tables in §2.1 ("Classification")), then the operators acting in it with Pauli matrices are conventionally denoted by τ_j. They are written in a basis where τ_3 is diagonalized and in the tables, I_3 denotes the eigenvalue $(1/2)\tau_3$ at the elements of this basis. However, sometimes the terminology "projection of the isospin on the third axis" is used for I_3; this is meaningless, because it bears no relationship to space-time. The eigenstates of τ_3 are identified as actual physical particles; the choice of the sign for the eigenvalue of τ_3 is determined by the electric charge of the particle (+1/2 corresponding to a particle with the greater charge; the ambiguity is resolved for K^o by taking into account the quark composition of the kaon). Similar remarks apply to the weak isospin space.

b. *The Gell-Mann Matrices*

The basis for *su*(3) is formed by the Gell-Mann matrices

$$\lambda_1 = \begin{pmatrix} 0 & 1 & 0 \\ 1 & 0 & 0 \\ 0 & 0 & 0 \end{pmatrix}, \quad \lambda_2 = \begin{pmatrix} 0 & -i & 0 \\ i & 0 & 0 \\ 0 & 0 & 0 \end{pmatrix}, \quad \lambda_3 = \begin{pmatrix} 1 & 0 & 0 \\ 0 & -1 & 0 \\ 0 & 0 & 0 \end{pmatrix},$$

$$\lambda_4 = \begin{pmatrix} 0 & 0 & 1 \\ 0 & 0 & 0 \\ 1 & 0 & 0 \end{pmatrix}, \quad \lambda_5 = \begin{pmatrix} 0 & 0 & -i \\ 0 & 0 & 0 \\ i & 0 & 0 \end{pmatrix}, \quad \lambda_6 = \begin{pmatrix} 0 & 0 & 0 \\ 0 & 0 & 1 \\ 1 & 0 & 0 \end{pmatrix},$$

$$\lambda_7 = \begin{pmatrix} 0 & 0 & 0 \\ 0 & 0 & -i \\ 0 & i & 0 \end{pmatrix}, \quad \lambda_8 = \frac{1}{\sqrt{3}}\begin{pmatrix} 1 & 0 & 0 \\ 0 & 1 & 0 \\ 0 & 0 & -2 \end{pmatrix}.$$

Acting upon the space of "flavours" in the space $\mathfrak{H} = 3_f$ generated by the quark states $q^\alpha = \begin{pmatrix} u \\ d \\ s \end{pmatrix}$, these operators are responsible for the approximate $SU(3)_f$-symmetry of the strong interaction. Another very important space with $SU(3)_c$-symmetry is the colour space.

The quantum numbers of the quarks *u*, *d*, *s* are the eigenvalues under the action on these states by the following operators (where λ_0 is the identity matrix):

$$I_3 = \frac{1}{2}\lambda_3 \,; \quad Q = \frac{1}{2}\lambda_3 + \frac{1}{2\sqrt{3}}\lambda_8 \,;$$

$$Y = \frac{1}{\sqrt{3}}\lambda_8 \,; \quad S = \frac{1}{\sqrt{3}}\lambda_8 - \frac{1}{3}\lambda_0 \,.$$

As we shall explain later, these operators also define the quantum numbers of the observable hadrons, but in different representations of *SU*(3).

OBSERVABLES

The general postulate concerning observables can be stated as follows. Corresponding to each quantity whose values can be measured on the states of a system with space $\mathfrak{H}$ we can associate a Hermitian operator A with the following properties.

a. The spectrum of A, that is, the set of real numbers a for which the operator $A-a$ is not invertible, is the complete set of values of the quantity that can be obtained by measuring the values at various states of the system.

b. If ψ is an eigenvector of the operator A with eigenvalue a, then on measuring A in state ψ we are sure to obtain a.

c. More generally, on measuring A in the state ψ, $|\psi| = 1$, we obtain a value in some interval (a, b) with probability equal to the square of the norm of the orthogonal projection of ψ on the eigenspace in $\mathfrak{H}$ corresponding to this interval: $<\psi \mid P^A_{(a,b)} \mid \psi>$.

In the previous subsection we indicated several operators corresponding to observables at internal degrees of freedom, namely, the projections of the spin and isospin, hypercharge, and strangeness. The quantum coordinates, the projections of the momentum, angular momentum, are operators on infinite-dimensional spaces and their spectrum can contain not only a discrete part, but also a continuous part and, finally, if the spectrum is unbounded, these operators are not defined at all vector states.

The operator $i\partial/\partial x^\mu$ converts the plane wave e^{-ikx} into $k_\mu e^{-ikx}$, consequently it corresponds to the projection of the 4-momentum onto the μth axis. On the other hand, $\partial/\partial x^\mu$ is one of the generators of the Poincaré group, namely a shift in the μth direction. Similarly, the infinitesimal spatial rotation reduces to the angular momentum observable.

EXPECTATION VALUES AND UNCERTAINTY RELATIONS

Let A be an observable, for simplicity, on a finite-dimensional space $\mathfrak{H}$. Clearly, $A = \sum a_i P_i$ where the (a_i) form the spectrum of A and P_i is the projection onto the eigenspace corresponding to a_i. For a normalized state ψ we have $<\psi \mid A \mid \psi> = <\psi \mid \sum a_i P_i \mid \psi> = \sum a_i \times${*the probability of obtaining* a_i *on measuring* A *in the state* ψ}. Therefore $<\psi \mid A \mid \psi>$ is the expectation value of A in the state ψ. In view of the statistical principles behind the description of interactions, such averages play an important role both in the mathematical formalism and in the interpretation of the theory. In quantum field theory, for example, they emerge in the form of expressions $<0 \mid \prod A_j(x_j) \mid 0>$, where $\mid 0>$ is the vacuum state vector, and $A_j(x_j)$ is the creation operator (of a one-particle state) of the field A_j at the point x_j of space-time. Experimental information can give the expectation values of A corresponding to supplementary averaging over certain states ψ, for example, over the polarization states if they are not fixed.

We denote by $A(\psi)$ the expectation value of A in the state ψ and by $[\Delta A(\psi)]^2$ the average value of the observable $[A - A(\psi)I]^2$ in the state ψ. In other words, $\Delta A(\psi)$ is the mean square deviation of the values of A from their expectation value. It is not difficult to prove the inequality

$$\Delta A(\psi).\Delta B(\psi) \geq \frac{1}{2} \mid <\psi \mid [A, B] \mid \psi> \mid ,$$

where A , B are two observables. This shows that the average spread of values of the non-commutating variables A, B, can not, in general, be arbitrarily small. In particular, among the observables of classical type there are *conjugate pairs* satisfying the relation $\frac{1}{i}[A, B] = 1$ (in Planck units $\hbar$). The pair (μth coordinate, projection of the 3-momentum on the μth axis) is a conjugate pair. It should be noted that the commutator of two operators can only be the identity in an infinite-dimensional space, so that at internal degrees of freedom such pairs cannot occur. For conjugate pairs we have $\Delta A \cdot \Delta B \geq 1/2$ independently of the state ψ. This is the Heisenberg uncertainty relation which in its time played an important role in the formulation of the ideas on how quantum mechanics agreed with classical mechanics in the boundary region of "semi-classical" phenomena.

For our purposes the role of the uncertainty relation is of importance in putting together the concept of virtual particles. In the calculation of such quantities as $<0 \mid \prod A_j(x_j) \mid 0>$ in QFT they are represented as series in perturbation theory, the terms of which correspond to Feynman diagrams similar to those depicted in §1. If the first few terms of the series give a good approximation to the experimental results, then it is reasonable to suppose that the corresponding process is in the main a superposition of quantum states of the field corresponding to these diagrams. By postulating that such states are real, we are forced to conclude in accordance with the rules for interpreting amplitudes, that these states are formed: a) by the asymptotically free states (plane waves) of the incoming and outgoing particles corresponding to the external lines of the diagram: b) by plane-wave states of particles corresponding to the internal lines for which the relativistic relation between energy and momentum breaks down: $k^2 \neq m^2$. Here the phrase "are formed " in the above sentence roughly means "is a tensor product", and the implicit one-particle states are space-time states. States of the field of type b) are called virtual particles, that is, "outside the mass surface". Their formal status is entirely clear. In order to argue the absence of a contradiction between the law of conservation of energy-momentum for real particles and the inequality $k^2 \neq m^2$ for virtual states, we invoke the Heisenberg uncertainty principle which is interpreted in the sense that in very small space-time regions, where the virtual particles are localized, there is a large degree of uncertainty in the conjugate momentum-energy variables.

THE HAMILTONIAN

The evolution of an isolated quantum system (or an isolated part of its degrees of freedom) with time can be described by supposing that the operator $U(t)$ taking $|\psi(0)>$ to $|\psi(t)>$ is linear (i.e., preserves the coefficients of a superposition of states) and unitary (i.e., preserves the amplitude of the transition between any pairs of states). Along with the condition for stationariness $U(t_1+t_2) = U(t_1)U(t_2)$, this enables one to conclude that $U(t) = e^{-itH}$, where H is an observable (Hermitian operator) called the Hamiltonian or the evolution operator. Here H has the dimension of energy and tH the dimension of action; action is measured in Planck constants $\hbar$ so that tH is a dimensionless quantity.

In the next section we shall explain how in QFT the basic problem is that of calculating the operator e^{-iS} and the expectations associated with it, where the action operator S is a functional of the creation and annihilation operators of the one-particle states of the fundamental fields.

In the case of a discrete spectrum, H can be diagonalized in the orthonormal basis $|\psi_j>$: if $H|\psi_j^{(k)}> = E_j|\psi_j^{(k)}>$, then $U(t)|\psi_j^{(k)}> = e^{-itE_j}|\psi_j^{(k)}>$, in other words, the states ψ_j are stationary, only their phases change. The set $\{E_j\}$ is the energy spectrum of the system. Let $\mathfrak{H} = \mathfrak{H}^{(1)}\oplus\ldots\oplus\mathfrak{H}^{(k)}$ be the decomposition into the sectors of states that are degenerate with respect to the energy.

The symmetry of the system with given H breaks from the full kinematic group$U(n)$ to the group $U(n_1)\times\ldots\times U(n_k)$ preserving the dynamics (commuting with H); this group consists of collections of separate unitary rotations in each sector. (The choice of U or SU, in view of our remarks about the phase, is related to the question of taking the interactions into account .)

SYMMETRY BREAKDOWN

We suppose that the Hamiltonian of some system is represented in the form $H = H_0+H_1+H_2$, where the system is totally degenerate with respect to H_0 (H_0 = multiplication by E_0), and H_1 and H_2 are small by comparison with H_0 (that is, their eigenvalues are much smaller than E_0). Let G_1 be the dynamical symmetry group of the Hamiltonian H_0+H_1; G_2, G_{12} are similarly defined. Then H defines the following hierarchy of breakdowns of symmetry:

$$\begin{array}{ccccc} & & G_1 & & \\ & \rotatebox{45}{$\supset$} & & \rotatebox{-45}{$\supset$} & \\ U(n) & & & & G_{12} \\ & \rotatebox{-45}{$\supset$} & & \rotatebox{45}{$\supset$} & \\ & & G_2 & & \end{array}$$

This hierarchy contains less information than the Hamiltonians themselves: it preserves the information on the sectors of degenerate states but forgets about the fine spectra. Since the time of the discovery of $SU(3)_f$-symmetry it has turned out that this package of data, namely the *hierarchy of symmetry breakdowns*, is a very successful intermediate stage on the route from the classification of experimental results to the choice of structures in quantum field theory corresponding to them.

When stationary states in $\mathfrak{H}$ are realized by physical particles, the energy spectrum is closely related to the mass spectrum of the particles, since in the rest frame of the plane wave, $k^2 = E^2 (= m^2)$. If we leave on one side the fact that quarks can only be asymptotically free at distances that are small by comparison with the hadrons and therefore we can only talk about their effective masses in different processes, we can consider from this point of view the breakdown of the symmetry $SU(3)_f$ acting in the space $\begin{pmatrix} u \\ d \\ s \end{pmatrix}$. It is violated because $m_u \approx 4$ MeV, $m_d \approx 7$ MeV, $m_s \approx 150$ MeV.

As has already been said, the nature of the interaction in QCD is such that the coloured particles turn out to be locked in a small volume with linear dimensions $\sim 10^{-13}$ cm. Furthermore, because of the uncertainty relation, massless quarks can have energy of order $(2 \times 10^{-11}$ MeVcm$)/10^{-13} \sim 200$ MeV. Accounting for a coefficient gives an energy of $\sim$ 400 GeV. For three quarks, the total energy is of the order 1200 GeV, which is close to the mass of the nucleon and accounts for its origins. It is clear that we can with great accuracy ignore m_u and m_d and regard the u and d quarks as being equivalent; this explains isospin symmetry. The accuracy of the $SU(3)_f$-symmetry, for which m_s has to be ignored is worse.

REPRESENTATIONS

We now recall that the groups $SU(2)_I$ and $SU(3)_f$ were originally discovered to be groups of approximate symmetries of observable states of elementary particles entering, for example, in the two octets of Table 1.

The action of $SU(3)_f$ on the 8-dimensional space induced, say, by the meson states is an example of a linear representation. The symmetry, exact or broken, corresponding to the group G or the diagram of its subgroups, can be realized on different spaces and an analysis, which can be purely mathematical in terms of the tensor algebra, leads to conclusions having an important physical interpretation. We now state several results on representations and give examples of their applications.

a. Suppose that the group G is represented by unitary operators in some Hilbert space $\mathfrak{H}$, that is, a correspondence $g \to T(g)$ is given, where $g \in G$, and $T(g)$ is an operator such that $T(g_1g_2) = T(g_1)T(g_2)$. We suppose that any subspace in $\mathfrak{H}$ that is invariant with respect to the action of all the operators $T(g)$ either coincides with the whole of $\mathfrak{H}$ or merely consists of the zero vector. We then say that the representation G in $\mathfrak{H}$ is irreducible. Example: the representation $S^{2J}(2)$ of $SU(2)$ (where 2 is the fundamental representation, $J = 0, 1/2, 1, 3/2, \ldots$, and S^{2J} is the symmetric power). Here is another example: the representations of $SU(3)$ of the form $S^a(3)$ and $S^b(\overline{3})$, where 3 is the fundamental representation, and $\overline{3}$ is its adjoint. A tensor product of representations, say, $T_1 \otimes T_2$, is the representation on $\mathfrak{H}_1 \otimes \mathfrak{H}_2$ by the operators $(T_1 \otimes T_2)(g) = T_1(g) \otimes T_2(g)$. The operators of the tensor power $T^{\otimes n}$ commute with symmetrization and antisymmetrization and therefore are defined on tensors of the appropriate symmetry.

b. Two representations of a group G are said to be equivalent if their spaces are related by a unitary isomorphism that commutes with the operators $T(g)$. For many groups, including the group $SU(n)$, it is possible to classify to within equivalence all the irreducible representations and to prove that any representation decomposes into a direct sum of mutually orthogonal irreducible sectors.

c. Irreducible representations can be described by various means. If we are interested primarily in the actual space of the representation and its tensorial make-up, then it is convenient to use a finite number of fundamental representations such that all the remaining ones lie in the tensor algebra of them. Thus, there is a realization of the representation $SU(3)_f$ in the space of the meson octet 8 in the form $\overline{3} \otimes 3 = 8 + 1$($\overline{1}$ corresponding to $u\overline{u} + d\overline{d} + s\overline{s}$), which is also an expression of the statement that the mesons consist of quarks and antiquarks.

If we want to give an effective description of the individual state vectors in the representation space, then we must choose eigenvectors for a maximal subgroup of the form $U(1) \times \ldots \times U(1)$ (or a system of commuting generators of it). The corresponding eigenvalues are quantum numbers of the states. Their discreteness is a direct consequence of the fact that all the irreducible representations of the group $U(1)$ are one-dimensional and are enumerated by the positive integers:

$$T_m(e^{i\omega}) = e^{im\omega}\,,\ m = 0, \pm 1, \pm 2, \dots .$$

The weight diagrams of Table 3 arrange the particles according to their quantum numbers, corresponding to the commuting generators Y, I_3, C of the Lie algebras $su(3)_f$ and $su(4)_f$.

We give a brief analysis from the above-described point of view of the scheme of the hypothetical $SU(5)$-unification of the fundamental particles and interactions. We shall write down explicitly only the degrees of freedom corresponding to the weak isospin and colour and restrict ourselves to the first generations, the remaining degrees of freedom being dealt with in similar vein. The left particles of the first generation are as follows:

$$\begin{pmatrix}\nu_e\\ e_-\end{pmatrix}_L,\ \begin{pmatrix}u\\ d\end{pmatrix}_L,\ e_L^+,\ \tilde{u}_L,\ \tilde{d}_L .$$

According to the above, their $SU(2)_W \times SU(3)_C$-composition has the form:

$$2_W \otimes 1_C + 2_W \otimes 3_C + 1_W \otimes 1_C + 1_W \otimes \bar{3}_C + 1_W \otimes \bar{3}_C .$$

We consider the group $SU(5) \supset SU(2)_W \times SU(3)_C$ with fundamental representation $5 = 2_W \otimes 1_C + 1_W \otimes 3_C$. We make the identifications:

$$\left\{\begin{pmatrix}\nu_e\\ e_-\end{pmatrix}_L, \tilde{d}_L\right\} = 2_W \otimes 1_C + 1_W \otimes \bar{3}_C = \bar{5} \quad (2_W \text{ and } \bar{2}_W \text{ are equivalent})$$

$$\left\{\begin{pmatrix}u\\ d\end{pmatrix}_L, e_L^+, \tilde{u}_L\right\} = 2_W \otimes 3_C + 1_W \otimes 1_C + 1_W \otimes \bar{3}_C .$$

The intermediate bosons carrying the $SU(5)$-interactions belong to a 24-dimensional subrepresentation in $\bar{5} \otimes 5$:

$$\bar{5} \otimes 5 \supset \underset{\text{gluons}}{1_W \otimes 8_C} + \underbrace{3_W \otimes 1_C + 1_W \otimes 1_C}_{\gamma,\, Z,\, W^\pm} + \underbrace{2_W \otimes 3_C + 2_W \otimes \bar{3}_C}_{X,\, Y,\, \tilde{X},\, \tilde{Y}}$$

DECAYS

The last theme of this section is the very useful phenomenological scheme describing the decay of quasi-stationary states in terms of the "non-Hermitian Hamiltonian". Such states of the field as short-lived elementary particles cannot strictly speaking be understood outside the interactions effecting decay. The simplest way of taking these interactions into account consists in postulating that the time evolution of the quasi-stationary state $|\psi>$ of the form

$$|\psi(t)> = |\psi(0)> e^{-i\tilde{E}t}, \text{ where } \tilde{E} = E_0 - \frac{1}{2} i\Gamma\,; \; E_0, \Gamma > 0$$

are real numbers, E_0 is the average energy of the state, and Γ is the *decay width* bringing about an exponential extinction of the state for a time proportional to Γ^{-1}.

The meaning of such a description can be made clear by trying to represent $|\psi(t)>$ for $t \geq 0$ in the form of a complex superposition of vectors $|\psi(0)> e^{-iEt}$ with different energies (masses) and interpret the coefficients of the representation $g(E)$ (that is, the Fourier transform of the function $|\psi(t)>$) as "the amplitude of the probability that ψ has energy E". The corresponding probability density, that is, the square of the modulus of the amplitude, then takes the form

$$P(E) = \frac{\Gamma}{2\pi} \frac{1}{(E - E_0)^2 + \Gamma^2/4} \,.$$

Thus in this scheme, the decaying state with average energy E_0 and width Γ is a continuous superposition of states with energy density as described above. The state itself can be associated with a pole of this density (or rather, the corresponding amplitude) in the complex energy plane; the residue of this pole (to within a factor) is the width, and the real part is the energy.

This last assertion about the relation between the physical states and poles of the amplitude has a very general significance and is widely applied in QFT. In general, the amplitude does not, of course, necessarily have such a simple (Breit-Wigner) form as in our model example.

§3.4. Quantization and Second Quantization

QUANTIZATION

The general scheme of quantum mechanics described above is given real substance if we have a means of explicitly constructing state spaces and operator observables that correspond to systems of interest to us.

At the time when quantum mechanics was being discovered, it was noticed that for a number of important systems (the atom as a system of several electrons in a Coulomb field; the electromagnetic field) the construction of the quantum description could be brought about in two stages: a) by the introduction of the corresponding classical Hamiltonian system; b) by the replacement of its classical observables by quantum ones (operators) by means of definite rules. This procedure has come to be known as quantization.

In the period of formation of quantum field theory it was discovered that the inclusion of quantum-field degrees of freedom that take into account the variability of the total number of particles, could be carried out by considering the previous particle wave functions as operators on a new state space, say, on the tensor algebra generated by the one-particle states, or more precisely, the part of it with the requisite symmetry conditions (Fock space). This procedure has acquired the name of second quantization.

HAMILTONIAN MECHANICS AND QUANTIZATION

In Newtonian mechanics, the motion of a point in a field of force is defined by the differential equation (*acceleration* = *force* × *mass*) and the initial conditions. The solar system is a model of a well isolated system of N points in a field of force created by their masses. Mathematically we can equivalently consider this as a single point in the $6N$-dimensional space of the coordinates and components of momenta of all the planets. It is called the phase space of the system.

The mathematical description of an isolated classical system with n degrees of freedom from which it is appropriate to transfer to a quantum description is, in fact, Hamiltonian mechanics. In Hamiltonian mechanics, three objects are associated with the system:

a. A phase space M of dimension $2n$ in which classes of local coordinates (q_i, p_i), $i = 1, \ldots, n$ are chosen; the so-called canonical coordinates.

b. The Poisson bracket, that is, an operation $\{f, g\}$ defined for any two observables (differentiable functions on M) in any canonical system of coordinates by the formula

$$\{f,g\} = \sum_{i=1}^{n} \left(\frac{\partial f}{\partial q_i} \frac{\partial g}{\partial p_i} - \frac{\partial f}{\partial p_i} \frac{\partial g}{\partial q_i} \right).$$

It is essential that $\{f,g\}$ should not depend on the coordinate system in which the function is calculated; this requirement can serve as a definition of the entire class of canonical coordinates if the Poisson bracket has been prescribed.

The phase space and the Poisson bracket defined on it completely determine the kinematic system. As in the quantum case, there is an important class of systems admitting a kinematic symmetry group G: Technically speaking, these are the orbits of G in its co-adjoint representation. The final object determines the dynamics.

c. The Hamiltonian H: the energy observable.

In order to write down the equations of motion it suffices to know the rate of change with time of any observable. This rate of change is defined by the Poisson bracket and the Hamiltonian according to the rule:

$$df/dt = \dot{f} = \{f, H\}$$

along any trajectory of the motion. We then obtain the following equations of motion in canonical coordinates:

$$\dot{q}_i = \frac{\partial H}{\partial p_i}, \qquad \dot{p}_i = -\frac{\partial H}{\partial q_i}.$$

It is clear here that the Hamiltonian H is constant along any trajectory: the evolution equation for H has the form $\dot{H} = \{H, H\}$. (Hamiltonian equations in which H depends on time are used to account approximately for the external effects on the system or the exchange of the energies with degrees of freedom such as heat dissipation, that are not accounted for.) Another important invariant that is conserved in the evolution is the phase volume of any region $U \subset M$ which is defined by the expression

$$\int_U dq_1 \ldots dq_n dp_1 \ldots dp_n ,$$

provided that U is completely covered by the canonical coordinates (q, p). This restriction on the surface of constant energy H = const, along which the classical motion proceeds, plays an important role in the statistical description of a classical system when the phase volume of the region is proportional to the probability of the state of the system being in it.

This role of the phase volume carries over to QFT, where in calculating the probability of transitions from the initial scattering state to the final one (say, of a system of plane waves), the classical degrees of freedom enter into this probability in the form of a factor proportional to the phase volume of the final states.

Suppose now that we are given a classical Hamiltonian system (M; *Poisson bracket*; H). By its quantization is meant the construction of a quantum system ($\mathfrak{H}$, $\hat{H}$) with the following rules of correspondence: there is a distinguished Lie algebra of observables on M (with respect to the Poisson bracket) containing H and a complete system of observables, as well as a representation of it by Hermitian operators in $\mathfrak{H}$, $f \to \hat{f}$, under which the Poisson bracket $\{f, g\}$ goes over to the quantum commutator $i[\hat{f}, \hat{g}]$.

All the known recipes are more or less refinements of the following classical rule. We choose a canonical system of coordinates (q, p) on M and we set :

$$\mathfrak{H} = \left\{\text{complex functions of } q \text{ with scalar product} \int \bar{f}g\right\} ;$$

$$\hat{q}_k = \text{multiplication by } q_k ;$$

$$\hat{p}_k = -i\,\partial/\partial q_k ;$$

$$\hat{H} = H(\hat{q}_1, \ldots, \hat{q}_n ; \hat{p}_1, \ldots, \hat{p}_n) .$$

The last expression is not in general uniquely defined because it is necessary to substitute non-commutating operators into the expression for H in place of the commutating coordinates. This difficulty does not arise for classical Hamiltonians of a point in a potential field: $H = T(p) + V(q)$, where T is a quadratic function of the momenta and V is the potential.

THE HARMONIC OSCILLATOR FROM THE POINT OF VIEW OF HAMILTONIAN MECHANICS

We consider the Hamiltonian system with $M = R^{2n}$, $H = T(p) + V(q)$, where T and V are quadratic functions and the "kinetic energy" T is positive definite. It is not difficult to prove that in a suitable system of canonical coordinates (which we denote by p, q as before) the Hamiltonian takes the form

$$H = \frac{1}{2}\sum_{i=1}^{n}\left(p_i^2/m_i + m_i\,\omega_i^2\,q_i^2\right).$$

This means that the system splits up into a direct sum of n independent "subsystems", namely the modes of vibration described by the Hamiltonians $H_i = \frac{1}{2}(p_i^2/m_i + m_i\,\omega_i^2\,q_i^2)$. Each mode is a one-dimensional harmonic oscillator (for $\omega_i^2 > 0$).

Both the classical and quantum motion of the harmonic oscillator lends itself to a full calculation and just for this reason the oscillator is an excellent model. But there is a deeper reason why the oscillator is ever-present. In mechanics the general potential motion with $H = T(p) + V(q)$ has equilibrium points : namely, the trajectories $(q^\circ, 0)$ for which $dV(q^\circ) = 0$. Near these points V has the form

$$V(q) = V(q^{\circ}) + \sum_{ij} \partial^2 V/\partial q_i \partial q_j \,(q_i - q_i^0)\,(q_j - q_j^0) +$$

$$+ (\textit{terms of higher order of smallness in } |q_i - q_i^0|).$$

If the quadratic form approximating V near q° is positive definite, then the trajectories of the oscillator corresponding to it are "small oscillations" of the system near a position of stable equilibrium.

In QFT this mechanism works in the calculation of the first quantum corrections to the classical motion of the system. Since the principle of action can be interpreted as the statement that the classical dynamics of a system is "the stationary value of it in space-time" (instead of the potential V one must work with the action S), the quantum fluctuations around the classical trajectory are described to a first approximation by a system of quantum oscillators.

We now give the main results of the quantization of the one-dimensional oscillator with $H = \frac{1}{2}(p^2/m + m\omega^2 q^2)$. The quantum Hamiltonian

$$\hat{H} = \frac{1}{2m}\,(-d^2/dq^2 + m^2\omega^2 q^2)$$

in the space of functions of q that decrease suitably rapidly, has the stationary state

$$\phi_n(q) = (m\omega/\pi)^{1/4}\,(2^n n!\,)^{-1/2}\,H_n(\sqrt{m\omega}\,q)\,e^{-m\omega q^2/2}\,,$$

$$H_n(x) = (-1)^n\,e^{x^2}\,d^n/dx^n\,(e^{-x^2}),$$

corresponding to the simple spectrum $E_n = (n+1/2)\omega$ (as always, the action is taken to be dimensionless by measuring it in Planck units $\hbar$).

FOCK SPACES

More generally, let $\mathfrak{H}$ be the state space of some system, which we will call a particle, and let us suppose that this particle is a boson or a fermion. This means that $S^m\mathfrak{H}$ (for bosons) and $\Lambda^m\mathfrak{H}$ (for fermions) are the state spaces of a system of m such particles, while the Fock spaces $S(\mathfrak{H})$ or $\Lambda(\mathfrak{H})$ (more precisely, certain completions) are the state spaces of an indeterminate number of such particles. Since particles are created and annihilated in the processes, one can attempt to describe the states of the quantum fields by vectors in such spaces or more generally, in spaces of the form

$$S(\mathfrak{H}_1^b) \otimes \ldots \otimes S(\mathfrak{H}_k^b \otimes \Lambda(\mathfrak{H}_1^f) \otimes \ldots \Lambda(\mathfrak{H}_l^f)$$

where $\mathfrak{H}_i^b$, $\mathfrak{H}_j^f$ are the spaces of single particle states of the different sorts of bosons and fermions laid down in the theory.

Let $\{\psi_1, \ldots, \psi_m, \ldots\}$ be an orthonormal basic system of states in $\mathfrak{H}$. We set

$$|\, a_1, \ldots, a_m\rangle = \begin{cases} \sqrt{(a_1+\ldots+a_m)!/(a_1!\ldots a_m!)}\; S\left(\psi_1^{\otimes a_1}\ldots\psi_m^{\otimes a_m}\right), \\ \sqrt{(a_1+\ldots+a_m)!}\; \Lambda\left(\psi_1^{\otimes a_1}\ldots\psi_m^{\otimes a_m}\right), \end{cases}$$

where the operators of symmetrization S and antisymmetrization Λ are to be applied respectively in the boson and fermion cases. The factors with factorials have been inserted in order to normalize the vectors $|\, a_1, \ldots, a_m\rangle$. Such a vector depicts the state of the complex system consisting of a_1 particles in state $\psi_1, \ldots, a_m$ particles in state ψ_m. In the fermion case, $a_i = 0$ or 1, in other words, antisymmetrization gives zero. The vector $|\, 0 \ldots 0 \ldots\rangle$ is called the vacuum vector. The states $|\, a_1 \ldots a_m \ldots\rangle$ form a complete orthonormal system.

The natural operators acting in the space of series are differential operators with respect to the independent variables with coefficients depending on these variables. Such operators are generated by multiplications by the variables ψ_i and the partial derivatives $\partial/\partial\psi_i$. Apart from this, it is customary to consider the creation operators $a(\psi_i)$ and annihilation operators $a^+(\psi_i)$ differing in their numerical coefficients. In the boson case they have the form :

$$a^+(\psi_i)\,|\, a_1\ldots a_i\ldots\rangle = \sqrt{a_i+1}\;|\, a_1\ldots a_i+1\ldots\rangle\,,$$

$$a(\psi_i)\,|\, a_1\ldots a_i\ldots\rangle = \sqrt{a_i}\;|\, a_1\ldots a_i-1\ldots\rangle\,.$$

In the fermion case analogous formulae are applied: one merely requires that the vector $|\, a_1 \ldots a_i \ldots\rangle$ be zero when at least one of the a_i differs from 0 or 1, and one has to insert the factor $(-1)^{a_1+\ldots+a_{i-1}}$.

For a single-particle state $\psi = \sum\lambda_i\psi_i$ we set

$$a(\psi) = \sum\lambda_i a(\psi_i),\; a^+(\psi) = \sum\bar{\lambda}_i a^+(\psi_i)\,.$$

The creation and annihilation operators $a^+(\psi)$ and $a(\psi)$ described by these formulae do not depend on the choice of the original basic system of states and satisfy the following system of commutation relations:

$$[a(\psi), a(\chi)]_\pm = [a^+(\psi), a^+(\chi)]_\pm = 0,$$

$$[a(\psi), a^+(\chi)]_\pm = \langle\chi\,|\,\psi\rangle$$

where $[A, B]_{\pm} = A\,B \pm B\,A$, and the plus sign refers to fermions and the minus to bosons.

In the second quantization formalism, the expression of all the objects in terms of annihilation and creation operators is a very convenient and physically informative form of notation. Instead of the state vector $| a_1 ... a_m ... >$ we can write $a^+(\psi_1)^{a_1} ... a^+(\psi_m)^{a_m} | 0>$ (to within a numerical factor). If (ψ_i) is a system of stationary single-particle states with energies E_i, then the Hamiltonian of a multiparticle system in the absence of interaction is simply given by $\sum E_i a^+(\psi_i) a(\psi_i)$. The interaction is taken into account by adding extra monomials of the creation and annihilation operators.

Continuing this line of argument, we can take the primitive object to be an operator algebra with the commutation relations written out, rather than the Fock space. In certain applications of these ideas it can be shown that this is equivalent to the Fock model, but in realistic models the position is more complex. Briefly it can be said that in QFT one calculates expressions of type

$$<0 | a^+(\psi_1) ... a^+(\psi_m) | 0>,$$

trying to manage without explicitly writing down the whole representation of the operator algebra and the functional-analytic context in which all such expressions would be formally well-defined mathematical objects. In most calculations of this sort, we begin by writing down formal expressions (divergent series of divergent integrals etc.) for the required quantities. These are then regularized by introducing extra parameters (cut-off momenta or distances, masses of massless particles, complex dimension of space-time) and studying how one forms the singularities of the quantities of interest to us at the physical values of the parameters. By the regularized value we mean, say, the term of the Laurent expansion in the parameters that follows the singular term.

In the next section we discuss the Lagrangian theories; these are the recipes according to which the information on one-particle states and symmetries is processed into formulae for the density of the action and interaction of the fundamental particles.

§4. The Lagrangian

§4.1. Action

We have already referred to the fact that among the trajectories of a Hamiltonian system with Hamiltonian $H = T(p) + V(q)$ there are positions of equilibrium, namely, points $(q^0, 0)$ of phase space that are stationary points of the potential energy observable. At these points, the difference $V(q^0 + dq^0) - V(q^0)$ has second order smallness with respect to dq^0, that is, $dV = 0$.

The content of the classical action principle for this model is that all the trajectories of the system, and not just the positions of equilibrium, are stationary points of a suitable functional. This functional is defined on the space of curves in phase space parametrized by time t with fixed end points:

$$\gamma = \{(q(t), p(t)) \mid q(t_0) = q^0, q(t_1) = q^1\}.$$

The value it takes at a curve is called the action along this curve and is defined by the integral

$$S(\gamma) = \int_\gamma pdq - Hdt = \int_{t_0}^{t_1} Ldt,$$

where L is the Lagrangian. The vanishing of the first variation of the action $\delta S = 0$ (under a variation of γ with the end points kept fixed) is equivalent to the Hamilton equation.

More generally we consider a classical system consisting of a collection of fields, say ϕ, in space-time . We suppose that on the kinematically possible values of the field in a region U of space-time we are given a functional $S(U, \phi)$ where the classical trajectories of the system are the extremals of this function: $\delta S = 0$. We can then say that the dynamics of the system is described by the variational principle. The functional S, being defined at all trajectories and not just those satisfying the equations of motion, is in general not uniquely determined by these equations. But quantum theory in which a number of the calculations can be interpreted as an averaging of $e^{iS(U,\phi)}$ over all the classical trajectories, shows that the rule for choosing the action functional is "more fundamental" than the equations themselves.

We now consider the general properties of the classical action functional which are also the primitives in QFT.

ADDITIVITY OVER SPACE-TIME REGIONS

If a region U is split up into a union $U_1 \cup U_2$ of disjoint sub-regions, then $S(U,\phi) = S(U_1,\phi) + S(U_2,\phi)$. This principle expresses a local property of the action functional. We split up the region into ever smaller pieces; if the value of the action in a small 4-volume becomes approximately proportional to this volume, then the action functional has a density $L(\phi)$ and can be written in the form

$$S(U,\phi) = \int_U L(\phi) d^4x.$$

The density $L(\phi)$ is called the Lagrangian. For the cases we are interested in, it is a function of the components of the fields and their space-time derivatives.

ADDITIVITY WITH RESPECT TO THE FIELDS

Suppose that a system of fields ϕ is split up into two subsystems: $\phi = (\phi_1, \phi_2)$. The action and its density can then often be naturally represented as a sum of three components:

$$L_1(\phi_1) + L_2(\phi_2) + L_{int}(\phi_1, \phi_2)$$

corresponding to the actions ϕ_1 and ϕ_2 in isolation and their interaction. If the L_{int} term is missing, then the fields ϕ_1 and ϕ_2 are classically completely independent: the dynamical equations for ϕ then take the form $\delta L_1 = 0$ and $\delta L_2 = 0$. In the general case these equations take the form $\delta L_1 + \delta_1 L_{int} = 0$ and $\delta L_2 + \delta_2 L_{int} = 0$, and the popular method of taking the interaction partially into account consists in regarding each of these equations in isolation ("particle in external field" and "field induced by a source").

In quantum field theory Lagrangians of free fields of matter and an electromagnetic field are quadratic with respect to the field and lead to linear equations of motion. But this is no longer the case for the Lagrangian of the free Yang-Mills field with non-abelian group responsible for the interaction. According to the general ideology passed down from quantum electrodynamics, the terms of the Lagrangian that are cubic and biquadratic with respect to the fields are regarded as a perturbation and are sometimes referred to as the "self-action".

SYMMETRIES

When the kinematic characteristics of all the fields that are to be taken into account are chosen and it is decided that the action functional is local, there is still a free choice for the Lagrangian. There are two further circles of ideas that restrict this freedom of choice:

a. Invariance with respect to the fundamental symmetries of the theory.

b. Renormalization.

Among the most important theoretical principles of symmetry that have been stated in the last two decades is the requirement that the Lagrangian be an invariant with respect to the gauge transformations connected with the internal symmetry groups. Gauge transformations bring about a rotation in the internal degrees of freedom, one at each point of space-time. They are described from a geometric point of view in the next section.

Renormalizability consists of the following property of the Lagrangian: in quantum-field calculations according to perturbation theory, ultraviolet divergences can be removed by carrying out a renormalization of a finite number of masses and coupling constants. We explain the formal aspect of this in more detail.

If the orders of the derivatives occurring in the Lagrangian are restricted (usually they are restricted to first order) and if it is required that they be of polynomial growth, then there remain only a finite number of independent possibilities for the choice of the Lagrangians of free fields and interactions. (Furthermore, symmetry requirements greatly reduce their number.) The free coefficients in the independent expressions of this kind are interpreted as priming masses and coupling constants measuring the interaction force. If they are not fixed by considerations of symmetry, then they must be introduced "by hand", in other words, from experiment. These constants are in general measured, and their physical dimension is determined by the fact that L is an action density.

By calculating from perturbation theory the amplitudes, cross sections, widths etc., we obtain a series in the coupling constants for which the coefficients of this series roughly speaking contain the powers of the energy of the virtual particles that are necessary to make the corresponding monomials dimensionless. The summation (superposition) is carried out over the energies of the virtual particles and if the dimension-cancelling powers of the energy are too large, the integral diverges at the upper limit: this is ultraviolet divergence. Renormalization is a regular procedure for removing these divergences (at least, at each term of the series, since the series as a whole is in practice always divergent and has to be regarded simply as an asymptotic series). The renormalizability requirement imposes a bound on the maximum possible number of degrees of the fields occurring in the Langrangian and sharply restricts the class of possibilities attainable.

For a long time this actual procedure of renormalization was regarded merely as a pragmatic recipe for extracting meaningful results from the infinite expressions. Nowadays renormalization, along with gauge invariance, have been made into a principle of choice of theories. This raises a number of questions on the physical meaning of this principle, of which we mention the following.

a. Renormalizability of modern gauge invariant theories with spontaneous breakdown of symmetry is usually proved via a special procedure, namely, dimensional regularization. As we have already mentioned, the dimension n of space-time is an artificially introduced variable parameter; at the physical value of $n = 4$, there occur singularities in the amplitudes. It would be a pity if this unusual situation proved to be merely an uninterpretable fragment of the formalism.

b. The problem of confinement in quantum chromodynamics has led to a number of approaches in which the non-local and non-perturbative nature of the gauge theory are emphasized. Does the renormalizability mean something beyond the bounds of perturbation theory ?

c. The search for all the new symmetry groups has led in the last decade to the following qualitative breakthrough, namely, the discovery of the fact that there exists a mathematical formulation of symmetry principles mixing up the field with the various spins and internal degrees of freedom with space-time. This occurs in supersymmetry and supergravitation models. Here, raising the symmetry is accompanied by a sharp decline in the divergences.

§4.2. Gauge Fields and Gauge Invariance

§4.2.1. FIELDS, FIBRE BUNDLES AND CONNECTIONS

Consider a fundamental particle, a lepton or a quark. As was explained above, associated with it is a Hilbert space of one-particle states, the elements of which can be written in the form, say, $q_{\alpha j}(x)$, where q is the symbol for the particle, x is a point of space-time, α is the colour index and j is the flavour of the quark. The quark field $q_{\alpha j}(x)$ at the point x takes values in the space of internal degrees of freedom. It can be a function, a generalized function or a generalized creation operator of the one-particle state. In this case, it acquires meaning after smoothing.

We draw attention to the fact that internal states of a particle at different points of space-time do not, in general, have to be identified in advance, since the postulate that such an identification is possible is a physical one rather than a mathematical one. In fact, in order to compare, say, the state of polarization of two particles, we must transfer them to the same point of space-time. But it may turn out that the result of such a comparison will depend on the history of the displacement, and then the absolute assertion that the original states are or are not the same is devoid of meaning.

This is in fact what turns out to be the case; say, in the Aaronov-Bom experiment an interference of two electron beams is observed going round a region of magnetic flux by different routes and this induces a relative rotation of the quantum phase.

The mathematical formulation of these ideas consists in the acceptance of the following three principles:

a. One-particle states of a material field are cross sections of vector bundles over space-time in which the fibre at each point is the internal space.

b. The fields, which are the carriers of the interactions, are the connections in these bundles.

c. The action is invariant with respect to the gauge group that rotates the degrees of freedom independently at each point.

We now clarify the concepts occurring here.

A bundle is a pair of differentiable manifolds E, M and a projection from one to the other $\pi{:}E \to M$. A cross-section of the bundle is a map $\psi{:}M \to E$ such that at each point x and M we have $\pi\psi(x) = x$, that is, the value of $\psi(x)$ lies in the fibre of π over the point x. In the cases of interest to us, M is space-time and the fibres of π are the internal degrees of freedom. The simplest example of a (fibre) bundle is the direct product $M \times F$, where F is a constant fibre. The cross sections of this fibre bundle are the functions on M with values in F. The best known physical example requiring the introduction of a non-trivial fibre bundle is the construction of the wave functions of a charged particle in a Dirac monopole field. The attempt to establish a general frame origin of the phase at all points of space-time (without a universal monopole line) leads to the appearance of discontinuities in the phase factors. These non-physical discontinuities can all be eliminated in an adequate treatment of the wave function as a cross section of a non-trivial fibre bundle. The physical effects of such global "topological degrees of freedom" in different situations have been all the better appreciated in recent years.

Let $E \to M$ be a fibre bundle. We suppose that for each arc γ going from x to y in space-time$M$ and for each initial internal state $\psi(x)$ of a particle at the point x we are given a final state $\psi(y)$ obtained as a result of the translation along γ. Such a set of parallel translations is called a connection in the bundle E. It is fitting for us to consider only those connections whose parallel translations are linear operators.

§4.2.2. POTENTIALS, FIELD STRENGTHS AND CURVATURE

Connections appear to be (and indeed are) different kinds of objects from the wave functions of matter. The standard method of approach for their description is to characterize a connection in terms of infinitesimally small translations, or what is the same, covariant derivatives. Covariant differentiation with respect to the coordinates x^μ is the operator $\nabla_\mu = \partial/\partial x^\mu + A_\mu$ acting on the cross sections of the bundle of matter:

$$\nabla_\mu \psi_k = \partial\psi_k/\partial x^\mu + A^j_{\mu k}\psi_j,$$

where j, k are internal indices. If the cross section satisfies the equation $\nabla\psi = 0$ along the curve γ, then it is said to be covariantly constant and the internal states corresponding to it are connected by a parallel translation.

The components of the matrix A_μ are called the potential fields of the connection. With respect to the Lorentz index μ, they behave like vector fields of matter, that is, fields of spin 1, and correspondingly the quanta of such fields are called vector bosons. But in this description, the difference between them and the fields of matter emerges formally in a second transformation law with respect to a change of coordinates in the internal spaces (passive point of view) or of the action of the gauge symmetry group (active point of view). Namely, if $\psi'(x) = U(x)\psi(x)$, then

$$\nabla_\mu\psi' = \partial_\mu\psi' + (UA_\mu U^{-1} + \partial_\mu U \cdot U^{-1})\psi',$$

that is,

$$A'_\mu = UA_\mu U^{-1} + \partial_\mu U \cdot U^{-1}.$$

Here the term $\partial_\mu U \cdot U^{-1}$, which takes into account the dependence of the gauge rotation on the point of space-time is essential; without it we would merely obtain an internal rotation in the space of matrices, or, technically speaking, in the adjoint group representation.

The above simple transformation law is obeyed by the field strengths:

$$F_{\mu\nu} = \partial_\mu A_\nu - \partial_\nu A_\mu + [A_\mu, A_\nu]\ ;\ \ F'_{\mu\nu} = UF_{\mu\nu}U^{-1}.$$

This can be verified formally, but it is better to satisfy oneself that $F_{\mu\nu}$ has a simple geometrical meaning; $1 + \varepsilon^2 F_{\mu\nu}$ is (to within ε^3) the matrix of the parallel translation along a closed loop represented by a parallelogram with sides ε parallel to $\partial/\partial x^{\mu}$ and $\partial/\partial x^{\nu}$. The non-vanishing of $F_{\mu\nu}$ means that there is curvature in the internal degrees of freedom.

As we shall see below, the field of the connection enters into the Lagrangian at terms of two kinds. The proper density of the action of the field A_{μ} is (to within a factor) the trace of the squared matrix $F_{\mu\nu}$ (averaged over $\mu\nu$) while the interaction with the matter is taken into account by means of terms with covariant derivative $\nabla_{\mu}\psi$.

§4.3. The Structure of the Lagrangian

THE FREE FIELDS OF MATTER

The classical Lagrangians for scalar (spin 0) and spinor (spin 1/2) fields of matter ϕ and ψ have the form

$$L(\phi) = \partial_{\mu}\phi^{+}\partial^{\mu}\phi - m^{2}\phi^{+}\phi\ ;$$

$$L(\psi) = \frac{i}{2}(\overline{\psi}\gamma^{\mu}\partial_{\mu}\psi - \partial_{\mu}\overline{\psi}\gamma^{\mu}\psi) - m\overline{\psi}\psi\ .$$

The coefficient m in front of the quadratic term in the fields is interpreted in both cases as the mass of the quantum of the respective free field. This is most simply explained by writing the equation of motion $\delta L = 0$ corresponding to its Lagrangians and substituting the plane wave ue^{ikx}. The relation $k^2 = m^2$ is obtained for the energy-momentum 4-vector.

The fields ϕ, ψ can have internal degrees of freedom. In this case, instead of the mass coefficient m we can write the general mass matrix, say, $\overline{\psi}M\psi$. The products occurring in the Lagrangian must then be interpreted as scalar products (convolutions with respect to the internal indices).

In what follows we shall encounter a situation in which terms of type $\psi^{+}H\psi$ enter into the Lagrangian, where H is the field with its dynamics, which has a non-trivial vacuum expectation value $\langle H \rangle$. This vacuum expectation value then turns up as the mass matrix for the original massless fields.

THE YANG-MILLS LAGRANGIAN

We write the covariant derivative in the form $\nabla_{\mu} = \partial_{\mu} - igA_{\mu}$, where A_{μ} is the field of the connection and g is the coupling constant (the meaning of this will be clear presently). The curvature form is then

$$F_{\mu\nu} = -ig(\partial_{\mu}A_{\nu} - \partial_{\nu}A_{\mu}) + g^{2}[A_{\mu},\ A_{\nu}].$$

In the standard models the matrix A_μ acts on the following degrees of freedom:

1_{em} (*phase factor*) in quantum electrodynamics: (A_μ) is then a photon.

$2_W \otimes 1_B$ (*weak isospin* $\otimes$ *B - field*) in the Salam-Weinberg electro-weak model; the (A_μ) are intermediate bosons and photons.

3_c (*colour*) in quantum chromodynamics; the (A_μ) are gluons. In particular, $[A_\mu, A_\nu] = 0$ for the group $U(1)$, say, $U(1)_{em}$, but $[A_\mu, A_\nu] \neq 0$ for non-abelian gauge groups. The Yang-Mills Lagrangian has the form

$$-\frac{1}{4g^2} F^a_{\mu\nu} F^{\mu\nu}_a$$

where a is the internal index.

The traditional method of quantizing a gauge field is to approach it in the same way that one approaches the vector field of matter. But here there arise difficulties, and to overcome them, a series of ad hoc methods has been developed. We give a brief description of the standard procedures.

a. The quadratic form in the derivatives, that is, the "kinetic" part of the Lagrangian $\frac{g^2}{4} G^a_{\mu\nu} G^{\mu\nu}_a$, $G_{\mu\nu} = \partial_\mu A_\nu - \partial_\nu A_\mu$, becomes, after taking the Fourier transform, a degenerate quadratic form which hinders the standard construction of propagators and the like. The reason for this degeneracy is clear: it is the presence of the gauge transformation group. Two fields related by a gauge transformation are physically equivalent if the symmetry group is not broken down. Therefore one can try to choose from each class of gauge equivalent fields one that satisfies suitable gauge conditions, such as $\nabla^\mu A_\mu = 0$.

b. In general, there are no terms in the Yang-Mills Lagrangian that are quadratic in A_μ. According to the standard wisdom, this means that A_μ describes massless vector bosons which do not exist in a free state in nature, apart from the photons. In the electro-weak model, the saving grace is the postulate that there are Higgs fields which "impart mass" to three of the four intermediate bosons by means of their vacuum expectation values. In chromodynamics we postulate that colour cannot escape; the gluons are locked in hadron bags and for the time being we do not have to worry about their masslessness.

c. In the Lagrangian there are terms of the third and fourth degrees in A_μ. They are treated as a self-action. Calculation from perturbation theory shows that in quantum mechanics the coupling constant g_s decreases with increasing momentum transfer: and this justifies the calculations of processes by perturbation theory "up to the boundary of the cell".

INTERACTION OF BOSONS WITH FERMIONS, CURRENTS

Even in the classical Dirac theory, the recipe was put forward of including this interaction: namely, to replace the ordinary derivative ∂_μ in the Lagrangian of matter by the covariant derivative ∇_μ. This leads to an interaction Lagrangian of the form $g\bar{\psi}\gamma^\mu A_\mu\psi$. In the abelian case, this can be expressed in the form gj^μ A_μ, where the current $j^\mu$is defined by the expression $\bar{\psi}\gamma^\mu\psi$. In the so-called V-A theory of the weak interaction, it was modelled on a Lagrangian that was represented as a product of two fermionic currents or a superposition of such products. This theory was not renormalizable but gave good results in a first approximation. In particular, weak interaction was characterized in it by a universal (that is, independent of the choice of the currents) dimensional Fermi constant $G \approx 10^{-5} m_p^{-2}$ (where m_p is the mass of the proton). In the present-day theory (or in its working version which operates with diagrams and virtual particles) the current emits a virtual intermediate boson which is absorbed by another current. To explain the high precision of the model with current-current interaction, the virtual bosons must be considered to possess mass, the exchange interaction must be assumed to occur at small distances and the theory must be made renormalizable. The Higgs fields ensure that this is possible.

HIGGS FIELDS, SPONTANEOUS BREAKDOWN OF SYMMETRY AND DYNAMICAL GENERATION OF MASSES

The Higgs field of general type is a scalar field of matter ϕ with internal symmetry groups G_H and a free Lagrangian of the form $\partial_\mu\phi^+\partial^\mu\phi - P(\phi)$, where P is a positive G_H-invariant polynomial, the minimum of which is attained not at $\phi = 0$, but on some non-trivial orbit of G_H. Classically it must be expected that the minimum of the potential energy P corresponds to a stable configuration. We suppose that in the quantum picture the vacuum expectation value $v =$ $= \langle 0 | \phi | 0 \rangle$ realizes this minimum and we introduce the new field $\chi = \phi - v$. The mass matrix for χ is the matrix of the second derivatives of P at the point χ. Let a be the dimension of the group G and b the dimension of the subgroup G_0 that takes v into itself. The mass matrix will then have $a-b$ zero eigenvalues corresponding to the directions of the translation of v by the symmetry group G_H. The corresponding "massless small oscillations" after quantization correspond to free massless Goldstone particles.

We now introduce the particles carrying the interaction corresponding to the group G_H, that is, a gauge fields A_μ . The kinetic term $\nabla_\mu\phi^+\nabla^\mu\phi$ gives the

Lagrangian of the interaction of the Higgs field with A_μ of the form $g^2\phi^+ A^+_\mu A^+ \phi$, which is quadratic in A and can for a constant field ϕ be regarded as the effective mass term for A (here g is the coupling constant). It turns out that it attaches non-zero masses to the $a-b$ components of the field A_μ, corresponding to exactly $a-b$ Goldstone bosons. The bosons themselves "disappear" as individual physical fields, since they can be used for fixing the gauge in the appropriate directions, and this turns them into longitudinal components of the gauge bosons. The gauge G_H-symmetry is spontaneously broken down by a reading of the vacuum expectation value; only the G_0-symmetry remains intact.

In the standard model of the electro-weak interaction the following selections are made: $G_H = SU(2)_W \times U(1)_B$, ϕ is a doublet with respect to weak isospin, ϕ_c is the charge-conjugate doublet; $P(\phi) = \frac{1}{2}\lambda^2(|\phi|^2 - \frac{1}{2}\eta^2)$. Let ϕ^+ be the charged component of ϕ and ϕ^o the neutral component; only one can have a non-zero vacuum expectation value by virtue of the conservation of charge. The covariant derivative is written in the form

$$\nabla_\mu = \partial_\mu - \frac{ig}{2}\tau_a A^a_\mu - \frac{ig'}{2} Y B_\mu ,$$

where τ_a acts on 2_W, Y is twice the average charge of the multiplet (the right hand components of the fundamental leptons e_R and ν_R are $SU(2)_W$-scalars, and the A_μ term is missing for these). The matrix of the effective masses for the A_1- and A_2-components has the form

$$\frac{\eta^2}{4}\begin{pmatrix} g^2 & 0 \\ 0 & g^2 \end{pmatrix},$$

and the matrix for the A_3- and B-components has the form

$$\frac{\eta^2}{4}\begin{pmatrix} g^2 & g'g \\ g'g & g'^2 \end{pmatrix}.$$

The zero direction for the latter matrix corresponds to a massless photon, while the direction orthogonal to it corresponds to a neutral intermediate boson Z; A_1 and A_2 are superpositions of W^+ and W^-. The Weinberg mixing angle is

$$\theta_W = \sin^{-1} g'/\sqrt{g^2 + g'^2} .$$

We note finally that by postulating the invariance of the total Lagrangian with respect to the gauge group of the weak isospin we have deprived ourselves of the possibility of introducing the standard mass terms of type $m\bar{\psi}\psi = m(\bar{e}_L e_R + \bar{e}_R e_L)$. Thus masses are also attached dynamically to the fermions in this model by the interaction with the Higgs fields. We write out the corresponding Lagrangians of the interaction for leptons of the same generation, say, e and $\nu = \nu_e$:

$$-f_e\left[(\bar{\nu}_L \bar{e}_L) e_R \begin{pmatrix}\phi^+\\ \phi^0\end{pmatrix}\right] + \{\textit{the complex-conjugate expression}\}$$

and the formula for ν is similar but with coupling constant f_ν.

We give a brief resume of the contents of this section. We listed a number of typical scalar expressions of fundamental fields that, in accordance with the dimensions of their coefficients, can enter into the classical Lagrangian. We also described certain working rules according to which information is put into or directly calculated from this Lagrangian without any actual quantum computations. These rules are based on analogies from quantum electro-dynamics, general quantum mechanical principles and a comparison with the previous more phenomenological models. In those cases where we came across a completely new phenomenon, such as confinement, these analogies did not provide anything and the Lagrangian of chromodynamics was chosen in the simplest form:

$$-\frac{1}{4} F^a_{\mu\nu} F^{\mu\nu}_a + \bar{\psi}(i\hat{D}_\mu + M)\psi,$$

where M is the mass matrix of the quarks. The electro-weak model gives more scope for imagination.

§4.4. From the Lagrangian to Reality

In this section we describe briefly how the theoretical model, embodied in the Lagrangian, can be compared with experiment.

We consider the Lagrangian of electro-weak interactions, more precisely the part of it relating to the lepton sector. In accordance with the analysis of the preceding section, we can extract from this Lagrangian the terms describing the amplitude of the interaction at one point of three particles, two leptons and an intermediate boson, that is, the vertices of the corresponding Feynman diagrams:

Interaction Lagrangian	Vertex
$-\frac{ig}{2}\bar{\psi}_\nu \gamma^\mu (\tau_1 A_{1\mu} + \tau_2 A_{2\mu})\psi_e$	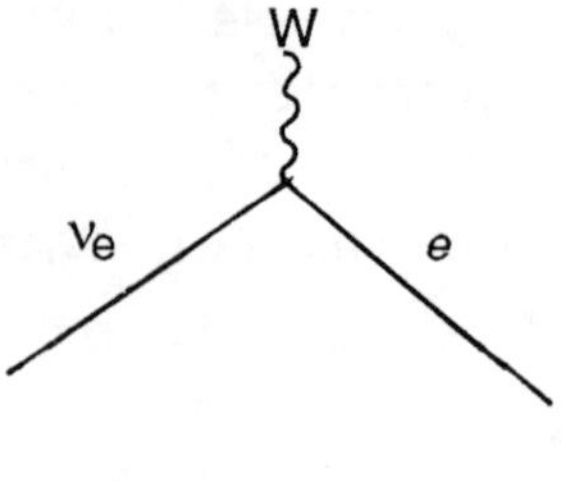

Consider the decay $\mu \to e + \overline{\nu}_e + \nu_\mu$. The simplest diagram that can give a contribution to the observable process must contain two vertices, and must have, for example, the form

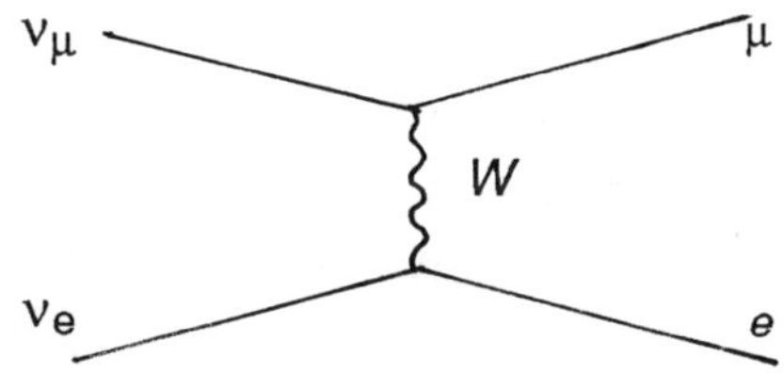

In good agreement with experiment is the supposition that such diagrams carry the dominant contribution to the process with the participation of four leptons and charge currents. More precisely, if we fix the wave functions of all the particles by assigning their momenta, internal and polarization states, we obtain a number, the amplitude of the corresponding process, after multiplying together the amplitudes of the vertices and the amplitude of the propagation of the intermediate particle. By taking the appropriate mean square of the moduli of these numbers over the unrecordable characteristics, we obtain probabilities which can be directly measured, for example, the width of the muon, that is, the total probability of its decay in unit time.

In the formula for the probability, the coupling constants, mixing angles, masses, and so on, enter as parameters. Therefore part of the experimental data has to be used for the determination of these parameters, and the rest for the confirmation of the predictions of the corresponding theoretical scheme.

In the characteristics for the lepton processes considered by us, the parameters occur in combinations $g^2/m_\pm^2$ and $\sin^2\theta$, where $m_\pm$ is the mass of $W_\pm$ and θ is the Weinberg angle. The mass of the intermediate boson appears in such a simple form if one uses the standard amplitude of the propagation in the form

$$-\frac{i}{\kappa^2 - m^2}(\eta_{ab} - \frac{k_a k_b}{m^2}),$$

which becomes $i\eta_{ab}/m^2$ under the hypothesis that the momentum transfer k_a is small by comparison with the mass.

Until the arrival of the theory of electro-weak interactions, the calculations of the amplitudes were carried out using the point four-fermion interaction

$$\frac{G}{\sqrt{2}}(\overline{\mu}\gamma^a(1-\gamma_5)\nu_\mu)\ (\overline{\nu}_e\gamma_a(1+\gamma^5)e).$$

The symbols of the particles denote their wave functions. Here G is the "old" weak interaction constant, for which one obtains the formula

$$\frac{G}{\sqrt{2}} = \frac{g^2}{8m_w^2}$$

in the more refined theory. On the other hand, the total width of the muon decay (after all the averagings have been done) takes on the form $\Gamma = G^2 m_\mu^5/(192\pi^3)$. Thus, by knowing g and m_W from the other experiments and by knowing the average lifetime of the muon, we can compare the calculated and observed values of Γ.

CHAPTER 3

Comments on the Dialogues

Comments on Dialogue 1

[1] The lecture given by A. Salam at the ceremonial session of UNESCO in honour of the centenary of Einstein's birth. The Russian translation of this lecture was published in the journal *Priroda*: A. SALAM, Einstein's final scheme: The unification of the fundamental interactions and properties of space-time" in *Priroda*, No. 1, 1981, 54-59.

[2] The lecture by S. Hawking read by him on the occasion of his appointment to the Lucasian Chair in Cambridge.

[3] These theories have been explicitly described in Ch. II. Apparently, there is the predominant opinion among physicists working in elementary particle theory that quantum chromodynamics and the theory of the electroweak interaction adequately describe the appropriate range of phenomena. Of course, it is not an easy task within the sociology of science to establish a consensus prevalent at a given moment. Phys has been called upon to be the vehicle for this, but the authors realize that it is difficult to avoid the danger of setting up their own evaluations as a consensus. With regard to the arguments in favour of the validity of QCD, see also the Comments to Dialogue 5.

[4] Phil is referring to the book by R. MILLIKAN, "Electrons (+ and -) protons, photons, neutrons, mesotrons and cosmic rays", University of Chicago Press, Chicago, 1947.

Millikan explicitly states that the new atomic physics is a realization of the programme of Democritus. "Democritus' principles, as quoted by Tyndall, are as follows:

1. From nothing comes nothing. Nothing that exists can be destroyed. All changes are due to the combination and separation of molecules.

2. Nothing happens by chance. Every occurrence has its cause from which it follows by necessity.

3. The only existing things are the atoms and empty space; all else is mere opinion.

4. The atoms are infinite in number and infinitely various in form; they strike together and the lateral motions and whirlings which thus arise are the beginnings of worlds.

5. The varieties of all things depend upon the varieties of their atoms in number, size, and aggregation.

6. The soul consists of fine, smooth, round atoms like those of fire. These are the most mobile of all. They interpenetrate the whole body and in their motions the phenomena of life arise.

These principles with a few modifications and omissions might almost pass muster today. The great advance which has been made in modern times is not so much in the conceptions and opinions themselves as in the kind of foundation upon which the conceptions rest. ....Today there is absolutely no philosophy in the field other than the atomic philosophy, at least among physicists."

[5] (W. Weber, Werke IV, 281), cited by MILLIKAN, ibid, p.20.

"The relation of the two particles as regards their motions is determined by the ratio of their masses e and e', on the assumption that in e and e' are included the masses of the ponderable atoms which are attached to the electrical atoms. Let e be the positive electrical particle. Let the negative be exactly equal and opposite and therefore denoted by $-e$ (instead of e'). But let a ponderable atom be attracted to the latter so that its mass is thereby so greatly increased as to make the mass of the positive particle vanishingly small in comparison. The particle $-e$ may then be thought of as at rest and the particle $+e$ as in motion about the particle $-e$. The two unlike particles in the condition described constitute then an Amperian molecular current."

[6] Phys is referring to Rutherford's classic article written in 1911. E. RUTHERFORD, "The scattering of α and β particles by matter and the structure of the atom", in Phil. Mag., **21**(1911), 669-688.

[7] Phys is referring to Millikan's book, ibid, p.13.

Information on the chemistry of the first half of the nineteenth century is contained in the book by V. KRASNOGOROV, *Yustus Libikh* (Justus Liebig), *Znanie*, Moscow, 1980, p.33: "The volume of the Annals of physics and chemistry for the year 1823 is open in front of me; here is published Liebig's memoir on fulminating combinations. Among the authors of the journal are Gay-Lussac, Tenard, Dulong, Laplace, Levi, Faraday, Metscherlich ..... Great names, classical ones, entered into the history of scientific research."

[9] Concerning the works of Faraday, see ibid p.47.

"In 1825 Faraday published his observations on the fact that hydrocarbons of the same composition (butylene and ethylene) can have different properties. Faraday suggested that as organic chemistry evolved, the number of cases of this type would increase."

[10] Concerning the papers of Lavoisier and Laplace, see Ya. G. DORFMAN's book *World history of physics from antiquity until the end of the 18th century*, Nauka, Moscow, 1974, 318 -326. In general it seems very likely that the chemical problem of establishing the true composition of chemical compounds and, connected with it, the problem of atomic weights, was the leitmotif of most of the investigations into the physics of gases, heat capacity of materials and problems associated with these themes. To evaluate the situation, articles have been written on general problems of chemistry in the encyclopedia of Brockhouse and Efron, written by D.I. MENDELEEV and his collaborators. See the articles "Matter", "Elements", "Chemistry", in the encyclopedia by F.A. Brockhouse, I.A. Efron, St. Petersburg, 1890-1967.

[11] See *Liste des Lauréats du Prix Nobel*, Almgrist and Wiksellers, Boktryckeri Actiebolag, Uppsala, 1967.

1908: M.E. Rutherford, professeur à l'université de Manchester pour ses recherches sur la désintégration des éléments et sur la chimie des corps radio-actifs."

The Nobel Prize was awarded to Marie Curie in 1911 (ibid).

[12] Concerning the Karlsruhe Congress, see D.I. Mendeleev's survey in D.I. MENDELEEV, *Khimicheskii kongress v Karlsruhe*, (Chemistry Congress at Karlsruhe), Collected articles, Vol. XV, Izdat. Akad. Nauk SSSR, Leningrad-Moscow, 1949, 154-165. Questions on the differences between atoms, molecules and chemical equivalents aroused such passions that it was decided to resort to a vote among the 150 chemists that had gathered at the Congress.

[13] See Millikan ibid p.22.

"For each chemical bond which is ruptured within an electrolyte a certain quantity of electricity traverses the electrolyte which is the same in all cases. This definite quantity of electricity I shall call *E*. If we make this our unit of electricity, we shall probably have made a very important step in our study of molecular phenomena."

[14] Apparently Maxwell was inclined to connect a current with a change in the state of the ether and not with the motion of a charge. In 1873 he wrote:

"It is extremely improbable that when we come to understand the true nature of electrolysis we shall retain in any form the theory of molecular charges, for then we shall have obtained a secure basis on which to form a true theory of electric currents and so become independent of these provisional hypotheses." See J.C. MAXWELL, *A Treatise on electricity and magnetism*, Vol. 1, Macmillan, Oxford, 1873 p.313.

[15] Math is reading a book "Geshichte der Physik", by M. von Laue, Bonn, 1950.

[16] There is a recent history of the discovery of the electron in the book: A.N. VYALTSEV, *Otkrytie elementarnykh chastits; elektron* β, *foton* γ, (The discovery of the elementary particles; the electron β, the photon γ), Nauka, Moscow, 1981.

[17] C.N. Yang, "Elementary particles. A short history of some discoveries in atomic physics".

"Thomson concluded that the cathode ray consisted of particles with a mass much smaller than the ions, and with a negative charge. He called them "corpuscles", and called their charge, which represented a basic unit of electric charge, the "electron". In later usage, however, the particles themselves came to be called electrons. Thus was born the first of the elementary particles recognized by man."

See ibid p.5.

"I cannot resist the temptation of showing you in the first figure a magnificent bust of the man who first opened the door to the physics of elementary particles."

[18] For example, in the book by D. KNIGHT, "Sources for History of Science", London, 1975.

"To put the question of the type "Who discovered chlorine or oxygen ?" is a hopeless business, only capable of confusing the question as to how did he attempt to become a scientist and why did his contemporaries show interest (or lack of interest) in this ?

[19] In Lorentz's book "Versuch einer Theorie der Electro-magnetischen und Optischen Erscheinungen in bewegenden Körpern", published in 1895 and summing up the research of the preceding period, Lorentz wrote "In the assumptions I shall be introducing there is a return, in a certain sense, to older ideas. One is not doing away with the essentials of Maxwell's ideas here, but it cannot be denied that the ions that I have introduced are not very different from the electric particles in use in earlier times."

See H.A. LORENTZ, "Collected Papers", Vol. V, Nijhoff, The Hague, 1937, p.8.

[20] Concerning Zeeman's work, see the books by von Laue and Vyaltsev.

[21] H. POINCARÉ, "L'état actuel et l'avenir de la physique mathématique", in Bull. des Sciences Math., Ser.2, **28**:1 (1904) 302-304.

[22] See the book G. HOLTON, "The Scientific Imagination", Cambridge University Press, 1978.

[23] One of the most accurate values was obtained by M. Planck in 1900 in the determination of the constant entering into the formula for black body radiation.

See M. PLANCK, "Uber eine Verbesserung der Wienschen Spectral-gleichung" in Vrh. Deutsch. Phys. Ges. **2** (1900) 202-204. The result was given in Planck's well-known book: M. PLANCK, "Vorlesungen über die Theorie der Wärmestrahlung", Barth, Leipzig, 1906.

[24] The tables referred to are updated once every two years by the group "Particle Data Group" in two versions. A complete version in the form "Review of Particle Properties", which is the April issue of the journal "Review of Modern Physics" . There is a short version "Particle Properties Data Booklet". The booklet is published by CERN, and on the cover is written "Available from Berkeley and CERN".

Comments on Dialogue 2

1) The notion of "formal reality" occurs in V.I. Vernadskii's manuscript "Outline of the history of the modern scientific outlook" based on lectures given at Moscow University in 1902-1903. The first three lectures were published in 1902 in the Journal "Questions of Philosophy and Psychology", SPB 1902 No.65. We have cited the publication given in the new edition "V.I. VERNADSKII, "*Izbrannye trudy po istorii nauki* (Collected works on the history of science)", Nauka, Moscow, 1981, p.38.

2) Phil was looking at two dictionaries: "A glossary of American technical linguistic usage" by E.P. HAMP, Utrecht, 1957. The Russian translation "*Slovar' amerikanskoi lingvisticheskoi terminologii* ", Progress, Moscow, 1964, pp.145-146.

The paradigm "the set of parallel primitive and resultant forms". "Paradigms can belong to different orders of abstraction; in other words, there exist paradigms of paradigms etc." "A set of contiguous words containing a common stem and all the affixes that can be adjoined to it form a paradigm." "One of a certain number of closed sets of words in each of which, words are contiguous but differ in their form and meaning, so that differences from set to set are parallel."

JEAN DUBOIS et al., Dictionnaire de linguistique. Librairie Larousse, Paris, 1973.

En grammaire traditionelle un paradigme est l'ensemble typique des formes fléchies que prend un morphème lexical combiné avec ses désinences casuelles (pour un nom, un pronom ou un adjectif) ou verbales (pour un verbe selon le type de rapport qu'il entretient avec les autres constituants de la phrase, selon le nombre, la personne, le temps: on dit *déclinaison* pour un nom, un pronom ou un adjectif et conjugaison pour un verbe. Ainsi, le paradigme de la première déclinaison latine est formée de l'ensemble des formes de rose <...>

2. En linguistique moderne, un paradigme est constitué par l'ensemble des unités entretenant entre elles un rapport virtuel de substitutabilité. F. de Saussure retient surtout le caractère virtuel de ces paradigmes. En effet, la

réalisation d'un terme (= sa formulation dans l'énoncé) exclut la réalisation concomitante des autres termes. A côté des rapports *in praesentia* le phénomène du langage implique également des rapports *in absentia*, virtuels. On dira ainsi que les unités *a,b,c* ... *n*, appartiennent au même paradigme, si elles sont susceptibles de se substituer les unes aux autres dans le même cadre typique (sintagme, phrase, morphème). Les paradigmes de flexion des langues <...> ne sont donc que des cas des rapports associatifs. La linguistique de F. de Saussure parlera d'une manière générale de *rapports paradigmatiques* là où le linguiste génevois parlait de rapports associatifs.

3) Course of Theoretical Physics: L. LANDAU and E. LIFSHITS, Vols.1-10, Nauka, Moscow, 1953-1973.

4) Among those physicists who consider that the description of the phenomena of consciousness lies *beyond* the limits of quantum theory is E. Wigner (see "Symmetries and Reflections", Indiana University Press, Bloomington-London, 1970).

5) The book referred to here is A. Sommerfeld's monograph which, in the early 20's, was a standard source of knowledge about the elementary particle physics of that time, namely, atoms, nuclei, protons, and electrons. We have cited the edition A. SOMMERFELD, *Atombau und Spectrallinien*, Friedr. Vieweg, Braunschweig, 1922, 8-9.

"Das Electron ist ein universelles Baustein aller Materie. Mag es im Electrischen Strome langsam dahinfliessen, oder mit aussertes Geschwindigkeit als katodenstrahl den Raum durcheilen ... may es den Gang des Lichtes in der Fernrohrlinze Beeinflussen, immer ist es dieselbe physikalische Einheit, seine Identität nachweisend durch gleiche Ladung und gleiche Masse, insbesonders durch gleiche Verhaltnis von Ladung und Masse. Wenn wir uns nach bisherigen Aussagen ein Bild des Electrons entwerfen wollen, so fällt dieses ziemlich dürftig aus. Im Grunde ist ein Electron, ebenso wie jede negative Ladung nicht anders als eine Stelle, in die electrische Kraftlinien von allen Seiten einmünden". It is then further explained that the electric field of the electron is completely symmetric in the rest frame of the electron which, by virtue of the theory of relativity, can always be constructed. There then follows an account of the difficulties encountered in the attempts to construct an extended electron. "Von der Angabe einer bestimmten Ausdehnung des Electrons sehen wir nach heutigen Vorstellungen besser ab." In conclusion Sommerfeld points out that the dimensions of the electron are in any case of order $< 10^{-5}$ times those of the atom.

6) What is being referred to here is that, according to the current theory of the electro-weak interaction due to Glashow, Weinberg and Salam, the

masses of the elementary particles arise as a result of their interaction with Higgs fields. Under spontaneous breakdown, these fields acquire non-zero values in the vacuum. At the early stages of the evolution of the universe, the higher temperatures give vanishing vacuum expectation values.
The articles are: D.A. KIRZHNITS, "The Weinberg model and the hot universe", *JETP Lett.* **15**(1972),745-748, and
D.A. KIRZHNITS and A.D. LINDER, "Macroscopic consequences of the Weinberg model", *Phys. Lett.* **42B** (1972),471-474.

7) In our mention of the ancient atomic theory, we have refrained from going into the details of its history; we have in mind the version of it set out in the book "De rerum natura" by Lucretius. This text was read and was influential in the establishment of the paradigm of the new natural science of the period of its formulation, which was in the 17th and beginning of the 18th centuries.

8) PHYS is referring to Vernadskii's book, *Loc. cit.*

9) So wrote Khvol'son in the book O.D. KHVOL'SON, "A course of physics", Vol.1, SPB (1902).

Comments on Dialogue 3

1) Michelson's first experiment in which an attempt was made to detect the motion of the earth relative to the ether, was performed in 1881 in Helmholtz's laboratory. The first of the articles of Lorentz relating to the problem of relativity appeared in 1886. Concerning this period of the development of the problem, see, for example, the article by I. Yu. KOBZAREV, "H. Poincaré's lecture and theoretical physics on the eve of the creation of relativity theory", Uspekhi Fiz. Nauk **11** (1974), 679-694.

2) R. Boskovich, of Croatian nationality (1711 - 1787) worked on astronomy and geodesy in Italy and France. In his book R.J. BOSKOVICH, "Theoria Philosophica Naturalis reducta ad unicum legem virium in existentium", Vienna, 1755 he discusses in detail the conjecture that matter consists of points of "primary elements" possessing mass but no extent and interacting in a universal fashion. The law of interaction was rather complicated: at infinity, the force became a universal force of attraction, while at small distances it changed its sign several times and became a repulsion. The complexity of this law of force gave rise to a hierarchy of aggregated corpuscles. Boskovich's theory can be regarded as a rationalization and systematization of the conjectures concerning the structure of matter as set forth in Newton's "Optics". Newton, on the other hand, was considerably less formally logical and also considered absolutely rigid particles in the style of the ancient atomists or ascribed to corpuscles ordinary properties of matter such as transparency.

For further information on Boskovich, see the encyclopedia by Brockhaus and Efron; also in "Contemporary history" by Dorfmann, pp.327-332 (see below).

The attempt to create a unified theory of matter undertaken by Boskovich attracted attention over a considerable period. References to him can be found in the works of many naturalists right up to the 19th century, from Faraday to Mendeleev. See, for example, Ya.G. DORFMAN, "A universal history of physics from the beginning of the 19th century to the beginning of the 20th", Nauka, Moscow, 1979, p.69; also the article by D.I. MENDELEEV, "Matter" in: "Encyclopedia", eds. F. Brockhaus, I.A. Efron, SPB, 1902,p.151.
"... it is useful to give by way of example the essential outlines of the teachings of Boskovich who nowadays is universally regarded as the founder, in a certain sense, of contemporary teaching on the properties of matter." To what extent

D.I. Mendeleev's statement corresponded to the actual consensus in 1902 is not clear to us, but it seemed reasonable to Mendeleev himself, since he was a thoroughly committed supporter of the Newtonian "paradigm" (see, for example, D. MENDELEEV "Two London lectures", SPB 1895).

3) The reference here is to the German physicist Abraham, author of a popular text book on electromagnetic theory at the beginning of the century. In Abraham's 1903 article "Prinzipien der Dynamik des Elektrons", *Ann. Phys.*, **10**(1903),105-179 the model of the rigid spherical electron is worked out in detail (both for a surface charge and a volume charge) under the hypothesis that the electron has no mass of its own. This model attracted considerable attention.

4) Concerning Sommerfeld, see Dialogue 4.

5) In the classical 1906 article, H. POINCARÉ, "Sur la dynamique de l'électron", reprinted in H. POINCARÉ, *Oeuvres*, vol.9, Gauthier-Villard, Paris, 1954, Poincaré calculated the action $\int dtdV\,(\mathbf{E}^2 - \mathbf{H}^2)/2$ for his model of the electron. Poincaré obtained the correct form of the Lagrangian function proportional to $\sqrt{1 - v^2}$ in contrast to Abraham's expression, which was not relativistically covariant; all the same, the sign he obtained for the Lagrangian was incorrect.

6) In the article M. PLANCK, "Das Prinzip der Relativität und die Grundgleichungen der Mechanik" in *Verh. Deutsch. Phys. Ges.* **8**(1906),136-141.

7) The question of the force of radiational friction is discussed in any good book on classical electrodynamics, for example, in the course by Landau and Lifschits, vol.2. L.D. LANDAU and E.M. LIFSCHITS, *Teoriya polya*, Nauka, Moscow, 1973. The history of the question of radiational friction could well be the subject of a separate study. The scientific literature of the 30's on this theme is apparently extensive.

8) Prout's articles were printed in the journal Annals of Philosophy in 1815-1816. Prout's starting-off point was that the atomic weights given in Dalton's book "Chemical Philosophy" were expressed in whole numbers. Dalton's whole numbers were, however, the results of the crudeness of measurements and rounding. Concerning Prout, see Yu.I. LESNEVSKII, "Antonius van den Broek", Nauka, Moscow, 1981.

9) Rutherford wrote about the mass defect as follows:

"Since experiments have shown that the nucleus has small dimensions, the positive and negative electrons that make it up must be densely packed. As Lorentz has shown, the electric mass of a system of charged particles depends on the interactions of their fields ... The packing has to be very dense in order that the change of mass created in this way be appreciable. This is how we can explain, for example, that the mass of the helium atom is not exactly four times the mass of the hydrogen atom." (ibid p.131).

10) A detailed analysis of the work of van den Broek, 1913-1914, dealing with these topics, is undertaken in the excellent study of Yu.I. Lesnevskii, op.cit., in which he gives a careful discussion of the role of this work in the context of the scientific developments of that period.

11) The reference here is to the two articles of Einstein: "Über einen der Erzengung und Vervandlung des Lichtes betreffenden heüristischen Gesicht punkt", *Ann.Phys.* **17**(1905),132-148 and "Zum gegenwärtigen Stand des Strahlungproblem", *Phys.Zeit.* **10**(1909),185-193.

12) A more detailed discussion on light quanta and the question of sources can be found in the article I. KOBZAREV, "A. Einstein, M. Planck and atomic theory" in *Priroda* No.3, 1979, pp.8-26.

13) The reference here is to the article A.H. COMPTON, "A quantum theory of the scattering of X-rays by light elements", *Phys.Rev.* **21**(1923), 207,483-572.

14) The history of the invention of Pauli's neutrino is set forth in his article "The old and new history of the neutrino" in the book W. PAULI "Aufsätze und Vortrage über Physik und Erkenntnistheorie", Braunschweig, 1961.

Fermi's papers appeared in 1933, 1934. The main publication is: E. FERMI "Versuch einer Theorie der β Strahle" *Zeitshr. für Phys.* **88**(1934),161-171.

According to Segre, Fermi originally did not like the method of second quantization but, as is clear from the above articles, necessity prevailed (see E. SEGRE "Enrico Fermi Physicist", Chicago, London, 1970).

"Fermi was a splendid figure in the problem of electromagnetic radiation. Initially he had some difficulties with the creation and annihilation operators introduced by Dirac, Jordan, Klein and Wigner, and in his first articles on quantum theory of radiation endeavoured to get away from these operators. Subsequently he completely familiarized himself with them and regarded the theory of beta decay as a suitable problem for practising this method." Segre goes on to write: "Fermi was fully aware of the importance of the success he had obtained and said that it was his best work and, in his opinion, he would be remembered for this research." (The reference cited here is the Russian translation: E. SEGRE, "Enrico Fermi Fizik", Mir, Moscow, 1973,pp. 99-101.)

15) Chadwick's work on the neutron appeared in 1932. The prehistory of the discovery of the neutron is somewhat confused. Many people had

"seen" it, but only Chadwick was able to give the unique interpretation of the experimental facts. See J. CHADWICK, "The existence of a neutron", Proc. Royal Soc. A**136**,692-708.

16) Reines's experiment is described in F. REINES and C. COWAN, "Free antineutrino absorption cross section ..."*Phys. Rev.***113**(1959),273-279.

17) Concerning Millikan's book, see Dialogue 1.

18) Dirac originally attempted to interpret his holes as protons, since the picture with a filled background seemed to be asymmetrical, admitting such a possibility. P.A.M. DIRAC, "A Theory of Electrons and Protons", Proc. Royal Soc. **126**(1930), 360-635.

This led to a contradiction since one then obtained the annihilations $e^{-} + + P \rightarrow \gamma + \gamma$, which meant that matter would become unstable. This was pointed out by R. Oppenheimer in the article J.R. OPPENHEIMER, "On the Theory of Electrons and Protons", *Phys. Rev* **35**(1930). In Pauli's survey which appeared in 1933, he refers to Oppenheimer and Dirac (without mentioning specific papers) and discusses the possibility of identifying the holes with particles with mass equal to the mass of the electron and opposite charge, but he writes :

"Dies erscheint schon deshalb unbefriedigend weil die Naturgesetze in diesen Theorie in bezug auf Electronen und Antielectronen exact symmetrisch sind. Sodann müssten jedoch (um die Erhaltungssatze von Energie und Impuls zu befriedigen mindestens zwei) γ-strahl Photonen sich von selbst in ein Electron und ein Antielectron umsetzen können. Wir glauben also nicht dass dieser Ausweg Ernstlich in Betracht gezugen werden kann."

W. PAULI, "Die allgemeinen Prinzipien der Wellenmechanik", in Handbuch der Physik B XXIV erste Teil, pp.83-272, see p.246.

The assertion that the masses of the particles and holes are equal is clearly stated in the article: P.A.M. DIRAC, "Quantized Singularities in the Electro-magnetic Field", *Proc. Royal Soc.* **133A**(1931),60-72.

A symmetric description of electrons and positrons appears in the article: W. HEISENBERG, "Bemerkungen zur Diracschen Theorie des Positron", *Zeitschr. für Phys.***90**(1934),209-231.

19) Anderson identified the positrons by examining the tracks of cosmic rays in a Wilson cloud chamber housed in a magnetic field, and determined the charge and mass of the positrons. Anderson made no mention of Dirac's theory and interpreted his observations by saying that he had discovered a new state of the proton with an increased radius equal to that of the electron. He considered everything in terms of a model of spherical particles with electro-

magnetic mass $e^2/2r$.
See C.D. ANDERSON,"The Positive Electron ", *Phys. Rev.* **43**(1933),491-494.

20) This is the article:W. HEISENBERG,"Über der Bau der Atomkerne I", *Zeitschr. für Phys.*,**77**(1932),1-11.

21) The conclusion that the nuclear forces for identical states of *PP* and *PN* are identical was made on the basis of an analysis of experimental data in an article by Breit and collaborators: G. BREIT, E.U. CONDON, and R.D. PRESENT,"Theory of Scattering of Protons by Protons", *Phys. Rev*, **50**(1936),825-845. A description in terms of an isotopically invariant nuclear potential is discussed in the articles: B. CASSEN and E.U. CONDON, "On Nuclear Forces",*Phys. Rev.* **50**(1936),846-849 and E. WIGNER, "On the Consequences of the Symmetry of the Nuclear Hamiltonian on the Spectroscopy of Nuclei",*Phys. Rev.* **51**(1936),106-119.

22) Yukawa's work appeared in 1935. Anderson's work on the discovery of the new heavy charged particle in cosmic radiation was published in 1936. In fact these were the μ-mesons. The story of the misadventures arising from the identification of the Anderson particle with the Yukawa meson is a very dramatic one. A brief account of this together with references to the original articles can be found in: K. NISHIJIMA, "Fundamental Particles", Benjamin, New York-Amsterdam, 1964.

23) Apparently the meson triplet was first introduced by Kemmer. A description of the results of the early period appears in the book: W. PAULI, "Meson Theory of Nuclear Forces", *Interscience*, New York, 1946.

24) W. PAULI op. cit.

25) The conclusion that there exists a heavy charged meson with mass approximately $1000m_e$ was made in an article by Leprince-Ringay and l'Héritier published in 1944. They studied cosmic rays. Apparently, this was a charged *K*-meson.

In 1947 a paper was published by Rochester and Butler who found two forks in a Wilson cloud chamber. One corresponded to the decay of a heavy neutral particle, and the other to a heavy charged particle. In the first case, they succeeded in determining the mass, which was roughly 550 MeV, and the decay scheme $\pi^+ + \pi^-$. Apparently this was the decay $K^0 \to \pi^+ + \pi^-$. The second case was not uniquely identified; very probably it was the decay $K^+ \to \pi^+ + \pi^0$. From 1951 on, work was done by a number of groups studying events in Wilson cloud chambers in showers caused by cosmic rays, and the disintegration properties of mesons and hyperons were discovered and studied; these are now said to be stable ($\tau = 10^{-8}$ sec - 10^{-10} sec). This stage of the development of elementary particle physics is well reflected in the series "Progress in cosmic ray physics" vols. 1-3, North Holland, Amsterdam, 1952-1954. Quantum field theory was, from the start, the main method of describing particles. The combination "Relativistic Field Theories of Elementary Particles" which we have now become accustomed to, was already in the title of a lecture by W. Pauli prepared for the Solvay Conference of 1939, which never took place: W. PAULI, "Relativistic Field Theories of Elementary Particles", *Rev. Mod. Phys.* **13**(1941),72-83.

In the paper by L. MICHEL, "Coupling properties of nucleons, mesons and leptons", *Prog. in cosmic rays*, vol.1,p.129, one reads:

"Many authors have attempted to develop fundamental theories giving the number of elementary particles, their rest masses, and so on ... Others have tried to describe some particles as constituted of several other "elementary" particles (generally a pair of fermions). However, up to now the only far reaching attempt for the description of the properties of particles is the field theory."

A brilliant survey of the theory of non-interacting fields was given by Pauli (op. cit.). The attempts to regard some of the elementary particles as being composite led to success in the end.

26) M. GELL-MANN, "Isotopic Spin and New Unstable Particles", *Phys. Rev.* **92**(1953), 833-834.

Gell-Mann remarked that by ascribing, say, to the particle Σ the isospin 1 (he combined Λ and $\Sigma^{\pm}$ into a single triplet), we forbid decays of type $\Lambda \to P\pi^-$ by preserving the isospin (and by the weaker condition that its T_3-projection is preserved). In similar fashion it was explained that $K \to 2\pi$ is forbidden. In 1953 Gell-Mann related all this to anomalous isospin (integral for Λ and half-integral for K). In fact, it then turned out that strong decay is also forbidden for Ξ, where the spin is the normal. Nowadays we regard the existence of the iso-invariance of strong interactions as accidental to some extent, while the fact that some transition is forbidden is related simply to the preservation of the separate types of quarks in the strong interactions. The difference in the values of T_3, say, in the reaction $\Lambda \to N\pi$ is, from our point of view, a reflection of two facts. 1) The quark composition of Λ is *uds*. 2) Thus for the quark compositions of $P \sim uud$ and $\pi^- \sim d\bar{u}$, we have *uds* on the left and *udd* on the right. The difference consists in the replacement of *s* by *d*, which stronger interactions cannot achieve. The $SU(2)$-isospin group itself is approximate. In this group, *d* is a doublet term and *s* a singlet, $T_3(d) = -1/2$, $T_3(s) = 0$, and $\Delta T_S \neq 0$. The change of the T_S-component of the isospin of the corresponding approximate symmetry $SU(2)$ serves as an indicator of the conversion $d \to s$, but this would be forbidden even in the absence of isotopic invariance as is indeed the case in decays of charmed particles.

The great success of the idea of isotopic invariance was that after all the failures of the description of the πN-scattering in the 0-300 MeV region within the framework of the field theory of the πN-interaction, a description of it was successfully obtained by assuming that the isobar dominates in it: that is, the resonance with $J = 3/2$ and $T = 3/2$. This explained the ratio of the cross sections at the peak $\sigma(\pi^+P \rightarrow \pi^+P)$: $\sigma(\pi^-P \rightarrow \pi^0N)$: $\sigma(\pi^-P \rightarrow \pi^-P) = 9{:}2{:}1$. This achievement precedes Gell-Mann's work. The conclusion that πN-scattering has a resonance character in the isobar region occurs in the note: K. BRUECKNER, "Meso-Nucleon Scattering and Nucleon Isobar", *Phys. Rev.* **86**(1952),626.

Compare this with the account at the beginning of §30 in the book: H.A. BETHE and F. de HOFFMAN, "Mesons and Fields", Vol.II Mesons, Harper Row, Peterson and Co., Evanston-New York, 1955,p.35: "As we have seen in the last section, there are so many points of disagreement between weak coupling theory and experiment that one should not put any faith in that theory ... Hence we are forced to consider a strong, rather than weak, interaction between the meson and nucleon ... For the present, we note merely that for strong coupling the theory is obviously more complicated, and it would be hopeless to develop any strong coupling theory, if it were not for some guiding principles. One of these is the well-known classification of quantum states according to angular momentum and parity. Even more fruitful has been the postulate of *charge independence*" (the italics are the author's). By the latter is meant isotopic invariance. Concerning confinement and groups, see Ch. II.

27) The work of Yang and Mills in which the gauge invariant equations for the isospin group were written down appeared in 1954: C.N. YANG and R.L. MILLS, "Conservation of Isotopic Spin and Isotopic Gauge Invariance", *Phys. Rev.***96**(1954),191-195.

The literature given in this work clearly shows how difficult it is to establish the real situation with regard to the origins of a paper by means of the references contained in it. The authors cite several fundamental papers that carried, from their point of view, the main contribution in the creation of isotopic invariance (restricting the discussion to nuclear forces), and after this they make the asertion that the π-meson has isotopic spin 1, with reference to Hildebrand's experiment on the measurement of the reactions $N + P \rightarrow \pi^0 + d$ and $P + P \rightarrow \pi^+ + d$. It is reasonable to suppose that the authors chose this work by virtue of the uniqueness of its interpretation.

A theoretical stimulus for the authors was undoubtedly the fact of the existence of local transformations of the phase of the wave function in electrodynamics, which was well known at that time. The authors are referring here to the well-known survey article of Pauli. Speaking a formal language, Yang and Mills looked for a formal generalization of this fact for the non-abelian isospin group.

Comments on Dialogue 4

1) W. Pauli regarded the quantum theory of the electromagnetic field as unsatisfactory since a) it did not explain the size of the fine structure constant α, and b) it led to divergences. He was expecting a subsequent profound change in the basic structure of the theory. See, for example, the article "On the idea of complementarity" 1948. (Russian translation in : W. PAULI, *Fizicheskie ocherki*, Nauka, Moscow, 1975, p.50, and especially p.56.)

"... All physicists agree that the current quantum theory which is inadequate for explaining the atomic nature of electricity and for the prediction of the values of the masses of the "elementary particles" in nature, can only have a limited domain of applicability ..."

A characteristic account of the attempts to remove the divergences in QED is given in the article by G. Wentzel written on the occasion of Pauli's 60th birthday: "All this activity surrounding the problem of self-energy has now been abandoned and is merely of historical interest. In spite of all its setbacks, quantum electrodynamics enjoys universal confidence as the best, albeit incomplete, instrument of the atomic theory". (G. WENTZEL, "Quantum field theory up to 1947", in : "Theoretical Physics in the twentieth century, a memorial volume to Wolfgang Pauli", ed. M. Fierz and V.F. Weiskopf, Cambridge, Mass. We have cited the Russian translation: *Teoreticheskaya fizika 20 veka*, Izdat. Inostr. Lit., Moscow, 1962, 60-93, see p.62.

2) We should restrict ourselves here to the statement of the question. The fate of the counter-programmes of the thirties and forties such as, for example, the programme of non-local field theories, and of the attempts to quantize space and other alternatives to the local field theory, and the evaluation of their importance and of the amount of attention given to them, should be the subject of a special study. Such a study is necessary for a history of science requiring an outline of the objective picture of scientific development, since in discarding the history of the counter-programmes developed at that time, we have inevitably made the development of it far more logical than was in fact the case.

3) A detailed history of the mechanical theory of heat can be found in the excellent book by Brush. (S. BRUSH, "The kind of motion we call Heat", North Holland, Amsterdam-New York, 1976.)

4) Phys is referring to the lecture given by Poincaré in St. Louis.

5) See, for example, Sommerfeld's speech at the Stuttgart Congress of Physicists in 1906. Sommerfeld expressed the hope that the Lorentz-Einstein relativity postulate would not be realized, since "the electrodynamic postulate" seemed natural to him: the model of the hard spherical electron was proposed by Abraham:

"I think we can say that physicists younger than 40 prefer the electrodynamical postulate, while those over 40 prefer the mechanico-relativistic." (Speech by Sommerfeld during the discussion of the lecture by M. Planck in Phys. Zeit. **7**(1906), 759-761.)

6) It seems that Phys has been reading Leibnitz:

"There are two sets of naturalists who nowadays make use of their fame and who derive their origins from antiquity: one of them has revived the teachings of Epicurus, and the other has, in essence, repeated the views of the stoics. The first considers every substance, including the soul and God himself, to be corporeal, in other words, consisting of matter or extended mass. Hence it follows that God cannot be omnipotent and omniscient; for how can a body influence everything, while in its turn it is not subject to the influence of anything and is indestructible ?"

(See *Leibnizens mathematische Schriften*, herausgegeben von C.J. Gerhardt, Bd. I-VII, Berlin-Halle, 1849-1863; Bd. VII,333, reference to G.W. LEIBNITZ, Collected Works, vol. 1, Mysl', Moscow, 1982,p.103.

Thus Leibnitz is ideologically against atoms. As a scholar he was not satisfied with ideological arguments and supplemented them with deductive inference from axioms in the booklet "The refutation of atoms obtained from the idea of contiguity of atoms" and the rider to it (loc. cit. Bd. VII, 284-288; loc. cit. Russian transl. pp.219-223.)

7) Phil's quotation is not quite correct. Here is the full text of the Blake poem:

Mock on, Mock on, Voltaire, Rousseau:
Mock on, Mock on; 'tis all in vain!
You throw the sand against the wind,
And the wind blows it back again.

And every sand becomes a Gem
Reflected in the beams divine;
Blown back they blind the mocking Eye
But still in Israel's path they shine.

The Atoms of Democritus
And Newton's particles of light
Are sands upon the Red sea shore,
Where Israel's tents do shine so bright.

(Quoted from *The Portable Blake*, Viking Press, Penguin 1977, p.142.)

8) V.I. LENIN, "Materialism and empirico-criticism",Zveno,Moscow,1909.

9) The question is discussed in detail in "The fundamental principles of statistical mechanics" in Brush's book (see 3)) . In Gibbs's preface to his book he writes: "Even if we restrict our attention to explicitly thermodynamical phenomena, we cannot avoid difficulties in such a simple question as the number of degrees of freedom of a diatomic gas. Of course, anybody who bases his research on hypotheses relating to the structure of matter, is building on an unsound foundation. Difficulties of this kind have kept the author from attempting to explain the mysteries of nature and have forced him to be content with the more modest problem of deriving certain more obvious principles relating to the statistical branch of mechanics."

We have quoted the Russian translation J.W.GIBBS, "The foundations of the principles of statistical mechanics", OGIZ, Moscow-Leningrad, 1946, p.14.

10) E. MACH, "Erkenntnis und Irrtum", Barth,Leipzig, 1905, Ch.I, §3.

"The representation of facts of reality in our thoughts or the adaptation of our thoughts to these facts enables us to supplement the facts intellectually merely by partial observation, since this supplementation is determined by the part observed ... This determination consists in the interdependence of the attributes of the facts which are the starting point for the thoughts. Since commonplace and immature scientific thinking is obliged to restrict itself to fairly crude adaptation of thoughts to facts, those thoughts that are applied to the facts are not always consistent with one another. Thus there emerges a new problem which has to be resolved to complete satisfaction by thinking: the problem of adapting one thought to another. This latter aspiration, causing a logical cleaning up of the thought process, but going much further than this goal, is a characteristic and advantageous sympton of scientific thinking, in contrast to ordinary thinking".

11) The problem of the quantization of the ion of the water molecule was solved by Pauli in 1922; see W.PAULI, "Uber das modell des Wasserstoff-molekulion" in : Ann. Phys. **68**(1922), 177-240.

12) Phys. is referring here to the works of Weinberg and Salam in which the definitive version was established of the renormalization of the Lagrangian of the weak interaction with Higgs expectation values. The correct form of the

vector multiplet was established earlier by Glashow in 1961. References to the literature were given in the commentary to Chapter II.

13) W. Heitler "The quantum theory of radiation", Clarendon Press, Oxford, 1936.

14) See, for example, the articles:

R.P. FEYNMAN, "The Theory of Positrons", Phys. Rev. **76**(1949), 749-759, and "Space-Time Approach to Quantum Electrodynamics", Phys. Rev. **76**(1949).

In Feynman's second article, in his discussion of the possibility of introducing cut-off functions into the integrals of QED so as to make the theory finite, he expresses doubts that this might make the theory contradictory. In discussing the possibility of replacing the ordinary propagator by the function f_{+}, which might make the answer finite, Feynman writes (ibid. p.778): "It could turn out that any correct form of f_{+}, which will guarantee energy conservation, may not at the same time be able to make the self-energy integral finite."

15) This stage of the development of the theory is very well reflected in the monograph by H.A. BETHE and F. de HOFFAN, "Mesons and Fields, vol.II Mesons", Harper Row, Peterson and Co., Evanston, Ill.-New York, 1955, pp.394-395.

"In this book we have treated the interactions of pions with nucleons, and that of nucleons with one another, as if the pion and nucleon fields formed a closed system, and as if no other particles existed. This approach is obviously open to question, and some physicists hold the view that no theory of either nuclear forces or the pion-nucleon interaction can be developed without including the curious particles. This would create a very difficult situation indeed, since the properties of the curious particles appear to be very complex and will take a long time to explore and even the number of such particles keeps increasing.

We see no reason for such a pessimistic view ..."

The authors then go on to give arguments in favour of the fact that virtual strange particles cannot give a large contribution to the nuclear forces. In fact, as it now turns out, they were correct; for the description of nuclei, it suffices to take into account the *u*, *d*-quarks and gluons. The contrast between the modern point of view and that held in 1955 is that we now know that one needs to use the QCD-interaction of quarks and gluons rather than the pseudo-scalar interaction of π-mesons and nucleons.

16) The correct form of the dispersion relations for scattering was given in the article M.L. GOLDBERGER, H.MYAZAVA, and E. OEHME, "Application

of Dispersion Relations to Pion-Nucleon Scattering", Phys. Rev. **99**(1955), 986-988.

Goldberger's proof was of pure heuristic character, which was exiguous from a mathematical point of view. The proof of the dispersion relation for πN-scattering in an interval of scattering angles was first given by N.N. Bogolyubov, B.V. Medvedev and M.K. Polivanov. It was reported at the International Conference in Seattle, September 1956 and published in the book: N.N. BOGOLYUBOV, B.V. MEDVEDEV, M.K. POLIVANOV: "Problems of the theory of dispersion relations", Gos. Izv. Fiz.-Mat. Lit., Moscow,1958.

17) The programme "Nuclear democracy" was most eloquently set out in the book G. CHEW, "Analytic S Matrix. A basis for nuclear democracy", W.A. Benjamin, New York-Amsterdam, 1966.

18) The quark model was proposed by Gell-Mann and independently by Zweig in 1964: M. GELL-MANN, "A Schematic Model of Baryons and Mesons", Phys. Lett. **8**(1964), 214-215; Zweig's article was published as a CERN preprint.

19) Such calculations can be found, for example, in the book A.I. AKHIEZER and V.B. BERESTETSKII "Quantum Electrodynamics", 3rd ed., Nauka, Moscow, 1969.

20) One of the most vivid documents relating to the discussions of that time can be found in the CERN preprint "Record of an informal discussion held on 7 July 1962", CERN 63-15, 1963.

21) This is the article: L.D. LANDAU, I.Ya. POMERANCHUK, "On the point interaction in quantum electrodynamics", Dokl. Akad. Nauk SSSR **102**(1955), 489-492.

22) This point of view was stated in articles by I.Ya. Pomeranchuk and co-authors in 1955. The reprint is in the book I.Ya. POMERANCHUK, "Collected Scientific Works. vol.II", Nauka, Moscow, 1972, pp.173-204. A typical expression of this occurs on page 38 (Work of 1955) in an article where it was proposed that the process $e^+e^- \to \mu^+\mu^-$ be investigated.

"If μ-mesons do not have some more important specific interaction than the electromagnetic one, then an experimental investigation of electrodynamical processes in which μ-mesons participate can provide important information on the limits of applicability of modern field theory and the nature of physical laws close to these limits, since the Compton wavelength of the μ-mesons is comparable with the dimensions where radical changes in the space-time concepts can be expected."

23) In the book W. HEISENBERG, "Physics and Philosophy", Harper, New York, 1958.

See also his book W. HEISENBERG, "Der Teil und das Ganze", Piper, Munich, 1971.

24) The research by Heisenberg on the non-linear spinor field is described in the book W. HEISENBERG, "Introduction to the Unified Theory of Elementary Particles", Wiley, London, 1966.

Comments on Dialogue 5

1) A survey on supergravity with a not too technical introduction: P. van NIEUWENHEUZEN, "Supergravity", *Phys. Rev.* No.4, 1981, 189-398.

2) Apparently, the best arguments in favour of the validity of QCD are: (1) Some quantitative results in the theory of jets, where results can be obtained by staying in the realm of ultraviolet freedom. (2) Results of calculations for the system $c\bar{c}$, $b\bar{b}$, obtained by a combination of perturbative QCD and a potential model for the systems $c\bar{c}$, $b\bar{b}$ themselves. Here the potential is chosen by interpolating between the action at small distances calculated in QCD in the framework of perturbation theory and close to the Coulomb potential (with logarithmic decreasing charge) and a linearly increasing potential that must correspond to confinement. The conjecture about this potential has a long history which we cannot go into here. At the moment we can apparently assume that it has been justified by computer calculations with a grid approximation to $\mathfrak{L}_{QCD}$.

There are also a large number of results obtained by supplementing QCD with ad hoc hypotheses and/or information taken from experiment.

The most accessible source for learning about the process of forming theories of elementary particles and the current state of the art is the Proceedings of the Conference on Elementary Particle Physics regularly held once every two years. The most recent: 1978 in Tokyo, 1980 in Batavia, U.S.A., 1982 in Paris.

3) Here again we run into homonyms. As in classical mechanics in its abstract form (the system of N particles with arbitrary pair potentials), we are not properly dealing with a "theory" in the sense that Phys was referring to earlier. Both classical mechanics and quantum field theory are a sort of grammar and dictionary of a specified type. Concrete physical theories are obtained by the choice of the potential and/or prescribing the system. In Newton's Principia in the famous Proposition XVI of section XI of the first book with its 22 corollaries, he gave the first sketch of this theory. This is the mechanics of a system of N bodies with pair potentials km_1m_2/r; which is the celestial mechanics of classical gravitation. In fact, the majority of the results obtained in this theory were obtained by passing to the problem of $N+1$ bodies where the fact that the mass of one of the bodies (the sun) is much larger than the mass of the planets is taken into account; and even more special problems are taken

into account. (For example, the theory of the system of the moon, earth and sun.)

Similarly, in "quantum field theory" we have some sort of general language. After choosing the specific candidates for the fields, one obtains concrete physical theories of type QCD or QFD. The situation is, in fact, more complicated. The majority of the realizations discussed in the scientific literature have a physically limited domain of applicability and mathematically can be contradictory; therefore, in practice, models are always supplemented with recipes enabling one to avoid the difficulties; for example, there is the method of renormalizations, which proves to be extremely effective in QED. The content of a book called "Quantum field theory" inevitably reflects both the state of the field of knowledge at the time of writing the book and the opinions of the author (or authors) on this topic. At any rate, in the last 20 years they have usually contained 1) classifications and a description of free fields; 2) a proof of a number of general theorems such as the CPT-theorem of Luders-Pauli; 3) a description of the construction of perturbation theory; 4) a description of the renormalization method; 5) a number of concrete models and the derivation of various results in these models.

The latest generation of books on field theory has tended to restrict the subject matter to gauge theories. In all cases this is justified by the fact that all present day theories claiming a wide domain of applicability rather than a narrowly restricted one, relate to the class of gauge theories. It can be hoped, apparently, that a theory such as QCD is in fact realistic and at the same time free of contradictions.

4) Phys goes on to talk about the standard rules of calculation for QFT in the concrete example of QED. An account of this and the derivation of these rules can be found in the books mentioned above. There is an extensive account of the computational results of QED in the monograph by Akhiezer and Berestetskii. A.I. AKHIEZER and V. BERESTETSKII, "Quantum electrodynamics", Nauka, Moscow, 1969.

Modern calculations within the framework of QED constitute a highly developed area. The accuracy attainable ($\sim 10^{-9}$) for such quantities as the magnetic moment of the electron is apparently unsurpassed in the history of science and exceeds the accuracy attained in celestial mechanics.

5) This, of course, was a programme rather than an already existing theory. See, for example, the title of the lecture of Einstein in Zurich "On the role of atomic theory in the latest physics". (See I. KOBZAREV, "A. Einstein, M. Planck on the atomic theory" in Priroda, No.3, 1979,pp.8-26, in particular, pp.8-9.)

The combination "Atomic physics" can be seen for example, in the title of the well-known text book: E. SHPOL'SKII, "Atomic physics", vol.1, 6th ed., Nauka, Moscow, 1974.

6) We are referring to the book by Heitler on quantum electrodynamics. Its first edition appeared in 1936, and we have cited below the text of the second edition, which appeared practically unaltered: W. HEITLER, "The quantum theory of radiation", O.U.P., Oxford, 1944. Apparently QED was studied at the beginning of the 30's most often from lectures given by Fermi in the U.S.A. in 1930: E. FERMI, "Quantum theory of radiation", in : Rev. Mod. Phys. **4**(1932), 87-132.

Comments on Dialogue 6

1) In what follows, Phil is referring to the articles "Theory" in two dictionaries : The Concise Oxford Dictionary of Current English, fourth ed., Clarendon Press, Oxford 1951; The American Heritage Dictionary of English Language, ed. William Morris, Am. Her. Publ. Co. and Houghton Mifflin Company, New York 1971. In what follows we refer to these articles as Oxf. Theory and Am. Her. Theory.

Oxf.; Theory. Supposition explaining something, esp. one based on principles independent of the phenomena etc. to be explained opp. to hypothesis, as: atomic theory, theory of gravitation, evolution; speculative view as, 'one of my pet theories' (often implying fancifulness); the sphere of speculative thought as "This is very well in theory but how will it work in practice ?, exposition of the principles of a science etc., as the theory of music; (Math) collection of results designed to illustrate principles of a subject, as theory of chance, of equations.

Am. Her.: Theory is a systematically organized knowledge applicable in a relatively wide variety of circumstances, especially a system of assumptions, accepted principles and rules of procedure devised to analyse, predict or otherwise explain the nature or behaviour of a specified set of phenomena. 1. Such knowledge or such a system distinguished from experiment or practice. 2. Abstract reasoning, speculation. 3. Broadly, hypothesis or suppositions. (Late Latin theoria, from Greek, contemplation, theory, from Theoros, spectator, from theas thai, to observe, from thea, a viewing. See theater.

2) This effect was worked out in the paper A.D. GALANIN and I.Ya. POMERANCHUK, "On the spectrum of μ-mesohydrogen", *Dokl. Akad. Nauk* SSSR **86**(1952),251-253.

Comments on Dialogue 7

1) Math is referring to the book T.S. KUHN, "The structure of scientific revolutions", Univ. of Chicago Press, Chicago, 1975.

2) Here Math is citing the famous "Cours de Linguistique Générale" by F. de Saussure. As is well known, this course was published posthumously from the notes of his students Bally and Sachet. The sense that Saussure himself invests in the terms *langue, langage* and *parole* is therefore not clear. Math thinks that this must be understood in the following way: 1. *Langue* is an abstract linguistic norm; it is some idealization of the norms of the language that actually operate; as for any idealization it does not properly "exist"; 2. *Langage* – a psycho-physiological structure that, so to speak, exists in the subconscious of the speaker and is responsible for the production of the spoken or written text; 3. *Parole* – the text itself. In principle, Math is not given to attaching much significance to the difference between *langue* and *langage;* in other words, he thinks that *langue* is in essence the "ideal" *langage*. Apparently he is also inclined to identify *langue* with the norm of the language in the form in which it is deliberately fixed and proclaimed by linguists in dictionaries and text books.

3) Math is referring to the book of one of the authors of the Dialogues: I. KOBZAREV, "Newton and his time", Znanie, Moscow, 1978.

4) See von Neumann's lecture "The Mathematician" in J.von NEUMANN, "Collected Works", vol.1, Pergamon Press, New York-London-Oxford-Paris, 1961,1-9.

5) Concerning Archimedes's statics and Ptolemy's experiments, see the book Ya.G. DORFMAN, "A universal history of physics from antiquity to the end of the 18th century", Nauka, Moscow, 1974.

6) The book referred to here is I. LAKATOS, "Proofs and refutations", Cambridge University Press, 1976.

7) "The theory of rigid bodies" has of course to be understood in the sense of "4" Oxford Dict. (See Comments on Dialogue 6). Of course, each theorist has his own different paradigm, an obvious analogue of the notion of "Idealect" in linguistics.

8) The book referred to here is "On the nature of things" by Lucretius.

9) This is a reference to an article written by Maxwell in 1859 in which he considers a kinetic model of a gas by using ideally elastic atoms.

10) The reference here is to the lecture given by Helmholtz in 1847 entitled "On the conservation of force". Information about this can be found, for example, in the book Ya.G. DORFMAN, "A universal history of physics from the beginning of the 19th to the middle of the 20th centuries", Nauka, Moscow, 1979.

Comments on Dialogue 8

1) The usual text books on quantum field theory, for example, the book: J.D. BJORKEN and S.D. DRELL, "Relativistic Quantum Fields, McGraw Hill, New York, 1965, contains brief summaries of the "Feynman rules". These rules lend themselves so easily to formalization that they can in principle be taught even to a computer; at least at the level of reducing the answer to integrals. The question on renormalization is more complex, but at least for quantum electrodynamics, they can be completely formalized.

The modern renormalization formalism for non-gauged fields was developed in papers by N.N.Bogolyubov and O.S.Pozasuk, and improved by R.Hepp. References to the original papers may be found in the book: N.N.BOGOLYUBOV and D.V.SHIRKOV, "Introduction to the Theory of Quantized Fields", Interscience, New York, 1959. Renormalization for gauge fields is described in the book: A.A.SLAVNOV and L.D.FADEEV, "Introduction to Quantum Theory of Gauge Fields", Nauka, Moscow, 1978.

2) Phys is referring to the book: R.P. FEYNMAN, "Theory of Fundamental Processes", W.A. Benjamin Inc., New York, 1961.

3) The work by Landau and Pomeranchuk was mentioned in Dialogue 4.

4) The Pauli-Villars regularization is a method of making all the terms of the perturbation series in QED finite by a formal modification of the Lagrangian. The method is convenient because the basic properties of the amplitudes and probabilities are not violated. After renormalization the usual answers occur. At energies $E > \Lambda$, however, the answers become meaningless because the condition that the probabilities of physical processes be positive is violated. A description of the method is given in the article W. PAULI and F. VILLARS, "On the invariant regularization in relativistic quantum theory", Review of Modern Physics **21**(1949),434-444.

5) T. KUHN, "Black-body theory and the quantum discontinuity, 1894-1912", New York, 1978.

Comments on Dialogue 9

1) According to the canonical statement of quantum mechanics, it is necessary to introduce a "classical apparatus" in order to include in the text of the theory, expressions of type "on measuring the quantity A one obtains the result A_i". An interaction with it is an act of measurement. It is postulated that the state of a "classical apparatus" is always ascertainable after the act of measurement. The act of ascertainment is not described within quantum mechanics; its realization leads to a sudden change in the "vector state" of completely finite type (reduction), although it is only statistically predicted. (The notion of a vector state is discussed below.) A detailed account of the theory of measurement can be found in the classical survey W. PAULI, "Die Algemeinen Principien der Wellenmechanik" in "Handbuch der Physik", 2nd ed., Vol.24, Part I, Springer, Berlin, 1933, 83-272.

Phys is referring to §7 of the classical monograph L. LANDAU and E. LIFSHITS, "Quantum Mechanics", Nauka, Moscow, 1974 as the standard version of the exposition of the theory of measurement. The account of the problems and controversies connected with the theory of measurement in quantum mechanics is not our concern here. For us the essential thing is that classical apparatuses cannot be deduced within quantum mechanics. In this sense classical mechanics is, in general, *not* contained in quantum mechanics.

2) The theory of the dispersion of light in matter as developed by Lorentz was based on the hypothesis that there existed elastically bonded electrons (oscillators). Lorentz's theory of electrons, based on a systematic development of this idea, was set out in the book H.A. LORENTZ, "The Theory of Electrons", Teubner, Leipzig 1916 (2nd ed.). References to the elastic oscillators can be found in this book. "This theory is highly reminiscent of the explanation put forward by various physicists working on the wave theory of light in its original form, when the ether was regarded as an elastic body. Zellmaier, Ketteler, Bussinek and Helmholtz showed that the speed of light must depend on the period of oscillation, since the body consists of small particles which are brought into vibration by the forces of the incident beam of light ... The small particles of Zellmaier have now become our electrons."

In the quantum theory of dispersion, the individual oscillators correspond to transitions from one level to another. See W. Heitler, op. cit, p.132.

"In fact, the first term of (12) corresponds exactly to the classical formula §5 (11), if we neglect the damping γ and attribute to each quantum transition of the atom a classical oscillator with the frequency $(E_i - E_0)/h$ and an 'oscillator strength' which is proportional to $P_{on,n_i} P_{n_i n_0}$."

3) In 1910 Debye applied Planck's formula for the energy of an oscillator to the normal modes of electromagentic vibration in a cavity and from this he deduced Planck's formula for the energy density of black-body radiation. The achievement here was that he avoided the problem of the interaction of material oscillators with the radiation; Planck had considered it within the classical theory which had lost its strength, as Einstein had emphasized in 1906. See P. DEBYE, "Der Warscheinlichkeit Begriff in der Theorie der Strahlung", Ann. Phys. **33**(1910), 1427-1434. A direct consideration of the proper modes of vibration eliminated this problem. The ideas of proper modes of a field were also applied by Rayleigh in 1900: LORD RAYLEIGH, "Remarks upon the Law of Complete Radiation", in: *Phil. Mag.***49**(1900),539-540. In fact, Planck's formula was applied even earlier in 1906 to oscillators in a cavity by Ehrenfest in the article P. EHRENFEST, "Zur Planckschen Strahlungtheorie", *Phys. Z.* **7**(1906),528-532; but this article remained unnoticed at the time, perhaps because the problem of the contradictory nature of Planck's deduction had not yet arisen. Somewhat later, similar ideas were expressed by Lorentz in correspondence with Planck. There is a discussion of the history of oscillators of this period in the article by I. Kobzarev that we have referred to and in the book: T.S. KUHN, "Black-body theory and quantum discontinuity", Oxford Univ. Press, New York, 1978,1894-1912.

4) The difficulties with the anomalous Zeeman effect in the classical theory of electrons were described in the book by Lorentz cited above. H. Poincaré discussed this question in the articles H. POINCARÉ, "La théorie de Lorentz et les expériences de Zeeman", in: *Éclairage Électrique* **11**(1897), 481-489 and "La théorie de Lorentz et le phénomène de Zeeman", ibid **19**(1899),5-15.

A correct quantum theory of the Zeeman effect requires that the electron spin be taken into account. An account of the history of its origins can be found, for example, in the article: B. van der WAERDEN, "The exclusion principle and spin" in: "Theoretical Physics in the Twentieth Century. A Memorial Volume to W. Pauli", ed. M. Fiez and V. Weiskopf.

5) Heisenberg's idea arose in the context of the old quantum theory of 1915-1925 in its direct application to atomic electrons and the radiation and scattering of light by atoms. Heisenberg made a radical change in the apparatus, and apparently this was motivated in the first instance by the need to bring

into the theory of dispersion of light the fact that resonances in scattering could only take place at frequencies of light equal to the difference in the energy levels of the atoms. The transition to tableaux, or in the language of our text, matrices of the coordinate operator, is based in his ideas on the "unobservability" of the trajectories in the atoms and on the analogy with Einstein's line of approach in his creation of the special theory of relativity. The removal of the "unobservables" can be regarded as the influence of the positivism of Mach but if this is the case, then it is indirectly through Einstein. See the book A. HERMANN, "W. Heisenberg", Rowlte, Hamburg, 1976, p.30, where the discussion between Kuhn and Heisenberg is quoted: "Haben Sie damals Mach gelesen ?

Answer: "Nein, Ich muss sagen dass ich niemals ganz ernsthaft Mach gelesen habe. Ich habe ihn später ein wenig studiert - viel später. Irgendwie war ich nie besonders von Mach beeindruckt. Ich war beeindruckt wie Einstein die Dinge auffasste."

The key point of Heisenberg's work of 1925 was the introduction of the relation equivalent to the commutation relation $[qp] = ih$ (although the exact form in which it was written is different). See the article: W. HEISENBERG, "Uber Quantentheoretische Umdeutung kinematischen und mechanischen Beziehungen", *Z. Phys.* **33** (1925), 879-893; equation (16). This equation was obtained by Heisenberg as a generalization (in the language of "tableaux" introduced by him) or the quantization matrices condition of the old quantum theory $\int pdq = nh$. The relation obtained by Heisenberg was already introduced by V. Kuhn and V. Thomas in the theory of dispersion. There are references to their work in Heisenberg's article.

6) The reference here is to the article: M. BORN and P. JORDAN, "Zur Quantenmechanik", *Z. Phys.* **34**(1925), 858-888.

In this article there is already a discussion of the quantization of the electromagnetic field. In the next article: M. BORN, W. HEISENBERG and P. JORDAN, "Zur Quantenmechanik II", *Z. Phys.* **35**(1926), 577-615, a quantum mechanics was constructed in matrix form for a system with n degrees of freedom.

7) E. FERMI: "Notes on Quantum Mechanics", University of Chicago Press, 1961.

We have cited the Russian translation E. FERMI, *Kvantovaya mechanika*, Mir, Moscow, 1965, pp.105-161.

8) A modern account of the second quantization procedure can be found in the book by Bjorken and Drell mentioned earlier.

9) The reference here is to the article: P.A.M. DIRAC, "Emission and

Absorption of Radiation", *Proc. Royal Soc.* A **114**(1927), 243-265. In Dirac's article there is a complete description of quantum mechanics for γ-quanta with a description of the amplitudes of the states with prescribed occupation numbers. The approach takes into account Bose-Einstein statistics, which was already known at that time, and uses quantization of oscillator fields. In fact, the latter had already been proposed by Born, Heisenberg and Jordan in the articles cited earlier; the main novelty of Dirac's work was, it would seem, the systematic treatment in the developed formalism of radiation and absorption of light by atoms.

10) The second quantization technique for particles subject to Fermi statistics was systematically developed in the article: P. JORDAN and E. WIGNER, "Über das Paulische Aquivalen verbot", *Z. Phys.* **47** (1928), 631-651.

In this article, the equivalence was shown of the description in the language of second quantization of the field and the occupation number , with the quantum mechanics of n particles in configuration space $\psi(x_1, ..., x_n)$ with antisymmetrized functions. The innovation was originally received without enthusiasm. See the survey (cited earlier) by Pauli, p.198:

"es ist zweifelhaft, ob es sich dabei um eine wirklich tiefgehende physikalische Analogie handelt, und es ist auch erwiesen, dass alle Resultate der Wellenmechanik auch ohne Anwendung dieser Methoden gewonned werden können. Zum mindesten als Rechenmethoden mussen sie aber angeführt werden". The method became basic for relativistic fields.

11) The method of Feynman diagrams can be found in any book on quantum field theory, for example in Bjorken and Drell's book.

12) In the article: N. BOHR and L. ROSENFELD, "Zur Frage der Messbarkeit der electromagnetischen Feldgrossen", *Kgl. Dans. Vid. Sels. Math.-Fys. Medd.* **12**(1933), 3-65.

An analysis was given of the precision with which the components of the electromagnetic field can be measured. It is shown that the maximal precision is determined by the commutation relations for the field.

13) Pauli's proof is set out in the article: W. PAULI, "The connection between Spin and Statistic, *Phys. Rev.* **58**(1940), 716-722.

14) M. BORN, "Quantenmechanik der Stossvorgänge, *Z. Phys.* **38**(1926), 803-827.

15) Phys is talking about Fermi's 1932 survey article, referred to above, where a systematic simple account is given of the theory of the interaction of atoms (atomic electrons) with radiation. An interesting feature of this article is

that Fermi does not make any use of second quantization. His field is decomposed into a sum of stationary wave oscillators; he then applies perturbation theory to the transitions of the atoms and oscillators from one state to another. This meant, in fact, that he was working in a basis where it was not the energy matrix of the field oscillators that was diagonal, but rather the amplitude of the oscillators. The matrix elements of the transitions for the operators turn out, of course, to be the same as the matrix operators of creation and absorption. The atom is described simply by the Schrödinger or Dirac equation. A number of problems that no longer appear in the standard text-books are considered in this article. For example, he considers the interaction of two atoms *A* and *B*, where *A* is excited at $t = 0$ and he shows that *B* starts to become excited only after time $t = r_{AB}/c$ with probability containing the factor $1/r_{AB}^2$. He considers the interference of the radiation from *A* with a wave reflected from a mirror and shows that the atom *B* "sees" the interference picture and the Doppler effect. He thus shows that quantum field theory reproduces all the expected features of classical wave theory. In the Dirac article referred to earlier, only the degrees of freedom of the electromagnetic field are discussed, that is, waves corresponding to photons. In Fermi's article, a systematic account was also given of Coulomb interactions; it is considerably simpler than that given in the earlier articles of Heisenberg and Pauli: W. HEISENBERG and W. PAULI, "Zur Quantenmechanik der Wellenfelder", *Z. Phys.* **56**(1929), 1-61; *ibid* **59**(1930), 168-190. The method developed by Fermi has been reproduced in a number of textbooks of the 30's and 40's (for example, in Heitler's book mentioned earlier).

In fact, even nowadays, the basis and Lagrangian are chosen when a concrete problem is considered, not from general rules of the paradigm of second quantization of fields, but according to the simplicity of the given problem.

Comments on Dialogue 10

1) See the book "Die Theorie der Strahlung und der Quanten", by A. Eucken and W. Knapp, Halle, 1914, p.365.
"Anderseits fiel mir bei den an dieser Stelle gehaltener Vorträgen und Diskussionen auf dass eine und dieselbe Theorie teils auf den Grundlagen der Alten Mechanik, teils aber auf solche Hypotesen aufgebaut wird, die zu dieser in Gegensatz stehen. Man muss hierbei in Auge behalten, das wohl jeder Satz ohne allzu grosse Mühe bewiesen werden kann, wenn man den Beweis auf zwei einander Wiederspechenden Prämissen begründet."

2) According to QED, the creation of virtual e^+e^-pairs by the Coulomb charge of the nucleus leads to a partial screening of nuclear charge by the distributed charge of the pairs that are close to the nucleus. This leads to noticeable effects for meso-atoms, as discussed in Dialogue 6.

3) V. S. POPOV, "Quantum electrodynamics of superstrong fields", *Priroda* No.10, 1981, 14-22, where references are given to the original research and survey articles.

4) The "non-classical character" of the physics of the twentieth century is tied up with the appearance of new relationships between what is being described and the description, which made themselves felt as a rift between one and the other. This rift is bridged by a philosophical thinking which redefines its theme to be not just reality or just an idealized objectivization of it in a theory, but the very premises of the possibility of such an objectivization and its meaning.

"Bohr's train of thought reminds me of the point of view expressed by Robert at the time of our trip to Lake Starnberger: atoms are in general not things", writes Heisenberg in his reconstruction of the state of mind of the physicists at the beginning of the 20's. "Although Bohr also reckoned that a large number of details of the internal structure of chemical atoms were known to him, the electrons that made up the shells of these atoms were in a clear sense no longer things. In any case they are no longer things in the sense of the earlier physics, namely, things that can be described without any reservations with regard to their position, velocity, energy and extent. I therefore asked Bohr: "If the internal structure of the atoms is so little capable of a clear-cut description, as you claim and if, properly speaking, you have no language in which you can talk about this structure, then how can we ever understand what an atom is ?" Bohr hesitated for a moment and then said: "Perhaps we can. But first we must find out what the word 'understanding' means ..." - and further on: "... a genuine new territory in some science can only be reached when at the decisive moment there is a readiness to abandon the description on which the earlier science rested, and in a certain sense to perform a leap into the void". (W. HEISENBERG, "Der Teil und das Ganze: Gespräche im Umkreis der Atomphysik", Piper, Munich, 1971, 63-64, 101.)

During the time that has elapsed since this Socratic dialogue, (..."The natural sciences rest on experiment and achieve completeness in the conversations of people who deal with them and discuss with one another the import of the experiment. Such conversations constitute the main content of a book. From this example it should be clear that science originates in dialogue. Furthermore, it is self-evident that after several decades, these conversations can no longer be reproduced verbatim" *loc. cit.* p.9), a language has been created in which it is possible to talk about the structure of atoms and of atomic and sub-atomic processes. This language which in its very essence is mathematical leads a double life since it has a double semantics. One of its faces is turned to a certain world of platonic entities which by the general concensus of mathematicians of the post-Cantor period is the receptacle of the meaning of all mathematical constructs. If one does not wish to appeal to this world, then one can at least say that the very tersely and precisely written language of set-theoretic mathematics is public domain and the criteria for its correct use do not arouse any discord. But as soon as the mathematical text becomes a discussion of theoretical physics, then it possesses a semantics turned towards physical reality and is interpreted according to different rules.

Furthermore, its mathematical aspect provides a strict system of limitations which determine both the evolution and the structure of the language (here it is not all that important whether special emphasis is laid on the syntactical, formal-algebraic structure of the computations or the analytical "rigour" and geometrical "clarity"). Reality is confronted with the discourse of theoretical physics. The formal-real opposition occurs at the junction: the incompatibility of the syntax of the language with that of the reality decribed in the preceding paradigm and the incompatibility of the two semantics. If the first was argued repeatedly and thoroughly, then the second was accepted as a necessity and, as it were, in desperation. The "void" between classical determinism and Born's probability is bridged by the dotted line of semi-classical approximations and estimates. But the "void" between classical probability and the quantum complex amplitude of the probability requires a leap even now.

Complex numbers, which play such a large part not only in the apparatus but also in the entire conceptual system of quantum mechanics, are worthy of great attention from this point of view. The opposition of the real and the formal is implanted even in the very etymology of the terms relating to complex numbers (real part, imaginary part). David Hilbert who was the first to put forward a clear-cut conception of the idea of existence in mathematics also used the illustration of $\sqrt{-1}$. In his celebrated "Mathematical problems" speech (1900) he says:

"In order to describe the significance of this problem (the determination of the consistency of the axioms of arithmetic) from yet another point of view, I should like to add the following remark. If some concept is endowed with properties that are mutually contradictory, then I say that this concept does not exist mathematically. Thus, for example, a real number whose square is equal to -1 does not exist mathematically. If, on the other hand, it can be proved that the properties possessed by some concept never lead via a finite number of inferences to a contradiction, then I say that the existence of this mathematical concept, for example, a number or a function satisfying specified conditions, is proved. In the case before us, where it is a question of the axioms of arithmetic of real numbers, a proof of the consistency of these axioms is equivalent to a proof of the mathematical existence of real numbers or the continuum. In fact, if one can completely succeed in proving the consistency of these axioms, then all those arguments that were at times brought to bear against the existence of the notion of real numbers, lose all foundation. True, from the aforementioned point of view, the notion of real numbers, or continuum, does not merely consist in the collection of all possible laws that can be obeyed by the elements of some fundamental series, but a system of elements the mutual relations between which are established by a system of axioms and for which the valid propositions are precisely those that can be obtained from these axioms by a finite number of logical inferences". (D. Hilbert, Mathematical problems II, quoted from "The Hilbert Problems", Nauka, Moscow, 1969, 26-29. The original can be found in *Göttinger Nachrichten*,1900, 253-297 or in *Ges. Abh. Werke*).

It is clear that the existence of a mathematical object in Hilbert's sense is in a different plane from the existence of an electron or a resonance. But this also is not without interest for modern physics, since the entire influential trend in the physics of the last two decades, which is axiomatic field theory, has set its sights on clearing up whether QFT exists in Hilbert's sense (if it is a question of interacting quantum fields). Here is not the place to touch upon the vulnerable aspects of the Hilbert conception, as regards its technical aspects, which were brought to light by Gödel. But the moral of Gödel's discovery is that the idea of existence cannot be replaced by that of formal consistency.

5) See "Die Theorie der Strahlung und der Quanten", p. 364. A further development of the adiabatic hypothesis can be found in the article: P. EHRENFEST, "A mechanical theorem of Boltzmann and its relation to the theory of energy quanta", *Proc. Amsterdam Acad.* **16**(1913), 591-597.
This hypothesis played a large part in the early stage of the development of quantum theory (1915-1925). Extra information on the early history of the adiabatic hypothesis can be found in Kuhn's book "Quantum Discontinuity", mentioned earlier.

6) For example, in the encyclopaedia by Brockhaus and Efron (quoted earlier) under the heading "Gravitation".

7) They are trying to recall the Hippocratic oath.

8) The reference here is to Poincaré's 1904 St. Louis lecture cited earlier.

Comments on Chapter 2.

The structure of elementary particle theory

Our text does not, of course, claim to serve as a textbook on quantum field theory or the physics of elementary particles. Its aim is to describe in mathematical language some of the basic structures of modern elementary particle theory.

The reader has to bear in mind that at the present time, our ability to tackle quantum field theory (or more precisely, some or other theory describing a specific type of interaction of elementary particles in the language of quantum field theory such as QCD or QFD) is to a large, even overwhelming, degree based on the use of the quantization of Bose and Fermi oscillators and the use of the Born approximation to perturbation theory. Of course, after we have, as a first step, made even the simplest Born approximation, the path is open to its improvement through the calculation of the radiation corrections which often presuppose a solution to non-trivial problems.

Thus we begin, as a rule, by excluding all the interactions between the fundamental fields contained in the Lagrangian. After this, the Hamiltonian of the system is converted into the Hamiltonian of a system of non-interacting oscillators corresponding to plane waves. For fields with spin 0 or 1, they are quantized in accordance with the Bose statistics while for fields with spin 1/2, the Fermi quantization is used. After this, standard procedures are used for constructing the S-matrix, that is, the transition amplitudes of processes of scattering type $\gamma+e\to\gamma+e$ or e^++e^- into $\gamma+\gamma$. Here we have in mind the usual construction for representing the interaction and the deduction of the Feynman rules, which is done by going over from the T-product to the N-product. We have not given an account of this traditional material, which can be found in the well-known textbooks, such as the books of Bjorken and Drell, or Akhiezer and Berestetski, mentioned earlier, or in the book N.N. BOGOLYUBOV and D.R. SHIRKOV, "Introduction to quantum field theory", 3rd ed., Nauka, Moscow, 1976.

One has only to recall that the basic aim for that period of the development of quantum field theory as reflected in these books, was considered to be the creation of methods going beyond the framework of perturbation theory; therefore the description, which is a relatively simple structure of second quantization and the passage to the Born approximation and the Feynman series, is not regarded in these books as the fundamental aim of the authors and can be interspersed in the description of more general and abstract methods. It goes without saying that even the ideal simplicity of the S-matrix and the Born approximation is rapidly demolished when one goes over to theories claiming to describe reality. Already in QED, the masslessness of the photon leads to the need for a special consideration of the infra-red catastrophe and special methods of constructing free states that take into account the presence of soft photons in physical states of electrons, and this, of course, can also be found in the above textbooks.

Even in QED, the standard canonical quantization procedure requires modifications relating to the presence of a gauge invariance in the theory; the quantization of the electrodynamical Lagrangian is in fact a quantization of a system with constraints. Technically, the question becomes more complex in the case of a gauge theory with a non-abelian group, but even here, as long as we restrict ourselves to perturbation theory, it would seem that no difficulties of principle arise. The contemporary state of the art is described in the books: A.A. SLAVNOV and L.D. FADDEEV, "Introduction to the quantum theory of gauge fields", Nauka, Moscow, 1978 and P. RAMOND, "Field theory, A modern primer", Benjamin-Cumming, London, 1981.

The use of gauge theory of QFD for the description of the weak interaction of leptons can again be based on the use of the Born approximation and it is fairly consistent in this sense. An example of a description of this type can be be found in the last section of part of Chapter 2. This theme has been considered in the book J.C. TAYLOR, "Gauge Theories of weak Interactions", Cambridge Univ. Press, London-New York-Melbourne, 1976 and L.B. OKUN', "Leptons and quarks", Nauka, Moscow, 1981.

Even in QED, when one has to consider bound states and calculate the annihilation of a positron into γ-quanta, one has to go beyond the bounds of the formal theory of the S-matrix and use the ψ-function of the bound state of a positron and a hypothesis on the factorization of large and small distances leading to proportionality of the probability of annihilation $|\psi(0)|^2$. In QCD, where the quarks are very strongly bounded, and in general they are not realized in free states, the problem acquires another order of complexity. All the same, the Born scheme is applicable in certain situations. In processes of type $\nu_\mu + A \to A' + \mu$, where A and A' are hadronic states and the transfer of energy and momentum to the hadrons is much larger than the characteristic energy Λ at which the QCD interactions become strong, the quark can be regarded as a point particle and one can use the Born approximation for the calculations. This leads to the theory of the so-called deeply inelastic processes which give a good description of experimental fact within the limits of its applicability. An extensive branch of QCD consists in calculations that reproduce the QED scheme for the positron, but what is not clear, of course, is the choice of the potential, or the factorizations of large and small distances, or the use of the parton approximation for gluons, for example, in calculating the process of annihilation of the heavy systems $\overline{q}q \to 3g$ (the g being gluons); this is far from obvious. Most likely, the time has not yet arrived for a systematic analysis of QCD.

APPENDIX

Strings

Recently I got an invitation to a conference on strings and superstrings to be held in El Escorial, Spain. A supplemented poster represented the famous San Lorenzo Monastery built by Felipe II as a Riemann surface of genus 17 with 4 cusps. The double symbolism of a monstrous castle-surface hanging by a cord was skilfully rendered by an artist. Rope and prongs were reminiscent of the Inquisition and the sadistic monarch being simultaneously a commonplace visualization of new toys of theoretical physicists - classical and quantum strings.

Well, not quite new. Foundations of string theory were laid down back in the sixties when Veneziano discovered his remarkable dual amplitude in strong interaction physics. It was soon understood that the Veneziano model describes quantum scattering of relativistic one-dimensional objects, i.e., strings, instead of common point-particles. This picture was in qualitative agreement with experimental data on the parton-like behaviour of strong interactions. One could imagine a meson as a tube of colour flux with quarks attached to its ends. The string length scale then should be roughly 10^{-13}cm. Since a string has internal excitation modes they could be held responsible for the proliferation of strongly interacting particles.

However the hadronic interpretation of dual string theory was plagued by many quantitative disagreements. To quote just one, it so happened that the quantum theory of relativistic strings seemed to be consistent only in 26-dimensional space-time while hadrons apparently lived in our 4-dimensional world !

Meanwhile quantum chromodynamics, i.e., the theory of quantized Yang-Mills fields, gained momentum as the right theory of strong interactions, and strings became outdated.

The modern Renaissance of the string theory is based first of all upon its reinterpretation suggested in 1974 by J. Scherk and John Schwarz.

The theory is now considered as a candidate theory of elementary particle physics at the Planck scale ($\sim 10^{-33}$cm.) rather than at hadronic scale. This romantic leap at twenty orders of magnitude from experimental data makes the situation in modern theoretical physics an extremely bizarre one and poses new problems of relating theory to the phenomenology of low energy (former high energy) physics. Psychologically this leap was prepared by a decade of Grand Unification models based upon Yang-Mills fields with a large gauge group and a bold extrapolation of the high-energy behaviour of coupling constants of strong and electro-weak interactions.

Another essential ingredient of the modern string theory, also created in the 1970's, is supersymmetry, i.e., a mathematical scheme which allows us to incorporate bosons and fermions in a multiplet mixed by a symmetry supergroup. At the classical level, this involves an exciting extension of differential and algebraic geometry, Lie group theory and calculus by the introduction of anticommuting coordinates which represent the half-integer spins of fermions. A string endowed with such fermionic coordinates is called a superstring. In a sense, supersymmetry implies general covariance and thereby requires unification with gravity.

The quantum theory of superstrings is consistent in 10-dimensional space-time. Since this is still far from our four dimensions, it was suggested, as a revival of an old Kaluza-Klein idea, that an extra six dimensions should be compactified at Planck scale. More specifically, our space-time presumably has a structure of a product $M^4 \times K^6$ where M^4 is the Minkowski space of special relativity while K^6 is a compact Riemannian space of diameter $\sim 10^{-33}$cm, i.e., a point for all practical purposes. For theoretical purposes, however, it is not a point by any means. In a fantastic paper [16], Ed Witten and collaborators proposed that (in a vacuum state) K^6 is a Calabi-Yau complex manifold with complicated topology which is responsible for such exotic properties of our Universe as the existence of three (or four) generations of elementary constituents of matter, i.e., leptons and quarks.

In the beginning of the eighties, Michael Green and John Schwarz discovered that requirements of consistent quantization (the so-called anomaly cancellation) place severe restrictions also on the possible gauge groups of the superstring theory. It seems now that a specific superstring model, called $E_8 \times E_8$ heterotic superstring, could eventually become a "Theory of Everything". Such are Great Physical Expectations.

Mathematically, the (super)string theory is no less interesting. As a young Moscow physicist Vadik Knizhnik once remarked, unification of interactions is achieved through unification of ideas. Physical papers devoted to many facets of string theory are now filled with homotopy groups, Kac-Moody algebras, moduli spaces, Hodge numbers, Jacobi-Macdonald identities and modular forms. A researcher trying to find his or her way in this mixture of seemingly disparate structures and techniques soon discovers that a physicist's intuition often transcends the purely mathematical one.

To this author, it was a challenging and highly stimulating experience.

SOME PHYSICS

Let me resume more systematically the physical content of modern quantum field theory before proceeding to its mathematical scheme.

In the 1920's, the fundamental physics consisted of four principal theories: electromagnetism; general relativity, i.e., gravity theory; quantum mechanics; and statistical physics. Broadly speaking, the first three dealt with "elementary" phenomena while the fourth one with "collective" ones and their general laws.

The scale of elementary phenomena was defined by four fundamental constants: e (electron charge), G (Newton constant), c (light velocity), $\hbar$ (Planck constant). The group of dimensions generated by them essentially coincides with the group generated by three classical physical observables: mass, length, time. In other words, starting from G, c, $\hbar$, one can define the "natural", or Planck, units:

$$M_{Pl} = (\hbar c G^{-1})^{1/2} \sim 10^{-5} g,$$

$$1_{Pl} = \hbar M_{Pl}^{-1} c^{-1} \sim 10^{-33} \mathrm{cm},$$

$$t_{Pl} = 1_{Pl} c^{-1} \sim 10^{-43} \mathrm{sec}.$$

The trouble is, we know of no elementary processes of Planck scale. Indeed, modern accelerators allow us to probe space-time at scales down to 10^{-16}cm and 10^{-26} sec only. On the other hand, M_{Pl} is the mass of a macroscopical drop of water about 0.2 mm diameter and elementary particles of such a large mass do not exist in our world.

This incompatibility of the three fundamental theories was long considered as evidence of the need of a deeper one combining all of them, in short, a $(G,c,\hbar)$-theory. In fact, two approximations to it had been discovered: general relativity which could be considered as a (G,c)-theory, and quantum electrodynamics, i.e., a $(c,\hbar)$-theory. Nobody as yet has succeeded in developing a consistent $(G,\hbar)$-theory, i.e., quantum gravity. The actual history of physics in our century followed an alternative course: since the discovery of radio-activity and subsequent construction of the first accelerators the list of elementary particles and forces tended to grow and immense efforts of several generations of physicists were devoted to the development of quantum field theory explaining the variety of observed phenomena.

In the 1960's we had to be content with the following picture. There are several sorts of matter particles, seemingly point-like ones, i.e., without discernible internal structure; stable matter consists of quarks and electrons. All matter particles are fermions, i.e., they obey Fermi statistics and have spin 1/2. There are also quanta of four fundamental forces: photons (electromagnetic force), gluons (strong force), vector mesons (weak force), graviton (?) (gravity). They are bosons, i.e., obey Bose-Einstein statistics and have spin 1 (or 2 for graviton).

Although elementary particles are point-like, they do have internal degrees of freedom. Mathematically, this means, in the first quantized picture, that a wave-function of a quark, say, is not a scalar function on the space-time but a section of a vector bundle associated with a principal G-bundle, where G is a Lie group called gauge group. (The choice of G ideally should have been governed by fundamental laws of nature but in the practice of the 1960's was model dependent). Similarly, a wave function of a quantum of fundamental force is a connection on the corresponding vector bundle, i.e., a matrix-valued differential form describing a parallel transport of internal state vectors along paths in space-time. A (second-quantized) theory of this kind is generally called a Yang-Mills theory.

The highest achievement of this epoch was (and still is) the so-called standard model describing electro-weak and strong interactions by means of the Yang-Mills fields with the gauge group $SU(3) \times SU(2) \times U(1)$, and several projects of Grand Unification based upon some large (preferably simple) group G containing $SU(3) \times SU(2) \times U(1)$. This large group should be a symmetry group of a fundamental theory at a higher energy, which is broken by some mechanism at lower energy leading to the effective Lagrangians of our present-day physics.

In all these developments, gravity could have been and had been neglected because the gravitational interaction between elementary particles was many orders of magnitude weaker than, say, the electromagnetic one (this is another way of saying that Planck mass is very large). In fact, the overall picture of our Universe is determined by different forces at different scales. At the scale about 10^{-13}cm, quarks are bound into protons and neutrons by the strong force. The atom nucleus consists of protons and neutrons bound by residual forces. The strong interaction, being a short-distanced one, dies out at atomic scale, and the electromagnetic interaction takes its part binding electrons and nuclei into a neutral atom. The electromagnetic interaction is long-distanced and very strong in comparison with gravity but for some reason, electric charges exist in both positive and negative kind which cancel each other very precisely in large lumps of matter like stars and planets. By contrast, gravitational charge, i.e., mass, never cancels, only adds up, so that at astronomical scale gravity becomes the main force. The residual electromagnetic forces, in the form of light and radio waves, serve as a source of energy and information for our kind of living matter.

(This hierarchy of scales reflected in the hierarchy of physical theories is a very characteristic trait of our modern understanding of Nature. Any future unified theory will have to explain it.)

Thus, all observed effects of gravity are in fact collective ones. They can become discernible at an elementary interaction level only in sufficiently excited matter, e.g., if an elementary particle is accelerated to the energy $\sim M_{Pl}c^2$ which is far outside the range of any conceivable installation. However, such conditions existed in the very early Universe, and the physics of these extremal states probably determined its later fate at cosmological scale.

Let us now reconsider on this background some features of string models.

Their first striking property is a prediction of a definite dimension of the space-time at Planck scale: 26 for bosonic strings and 10 for superstrings. This source of embarassment in dual models of hadrons now becomes one of the major predictions of the theory. However, it is not directly testable and it poses the problem of explaining the apparent four-dimensionality of the low-energy world. Very loosely speaking, one can imagine that $26-10 = 16$ dimensions somehow conspire to accommodate the internal degrees of freedom of fundamental particles (16 being the rank of the gauge groups $E_8 \times E_8$ and $SO(32)$), while the remaining $10-4 = 6$ dimensions "spontaneously" compactify at the Planck scale early in the course of cosmological evolution.

Below we shall have more to say about the mathematical origin of these critical dimensions, 26 and 10. It suffices to mention here that their appearance is a pure quantum field theoretical effect.

The second property of string models is the unification of the four known forces, including gravity, in an effective Lagrangian of low-energy approximation to the full theory which itself is much richer.

Their third property is a prediction of (very tight constraints on) the possible Grand Unification gauge group. It may be $E_8 \times E_8$, one factor being responsible for common matter while another one for the so-called "dark matter" interacting with the common one only through gravity.

Their fourth property is the incorporation of supersymmetry into the framework of fundamental physics.

One must add that such a theory actually does not exist yet. It is rather an ideal picture, a puzzle some of whose fragments have marvellously found their proper places, while others still remain a challenge.

Besides, all of this may prove some day to be just wishful thinking — as physics. Fortunately, mathematics is less perishable.

MATHEMATICAL STRUCTURE OF QUANTUM FIELD THEORY

In the following pages I shall try to make explicit some of the mathematical structure of fundamental physics stressing specific properties of string models.

a. *Virtual classical paths and action.*

A model of a physical system starts with a description of a set $\mathfrak{P}$ of "virtual classical paths" and an action functional $S: \mathfrak{P} \to \mathbf{R}$. Generally speaking, $\mathfrak{P}$ is a function space, e.g., a space of maps of manifolds $f: M \to N$, or a product of such spaces. Maps may be subject to certain boundary conditions; N may be a bundle over M and $\mathfrak{P}$ may consist of sections of this bundle, etc. Usually M and/or N are (pseudo-)Riemannian manifolds (with fixed or variable metrics) and S is obtained by integrating a natural volume form over M or N. Examples:

General relativity. M is a fixed 4-dimensional C^∞-manifold, $\mathfrak{P}$ is the space of Lorentz metrics $g = g_{ab}dx^a dx^b$ on M (i.e., sections of $S^2(TM) \to M$ with positivity conditions),

$$S(M, g) = -(16\pi G)^{-1} \int_M R \, \mathrm{vol}_g \quad \text{(Hilbert-Einstein action)} \tag{1}$$

where G is the Newton constant, R = Ricci curvature, vol_g = the volume form of g.

Massive point-particle propagating in a space-time (M, g). Here $\mathfrak{P}$ = = maps: $\gamma:[0,1] \to M$,

$$S(\gamma) = -m \int_0^1 ds, \tag{2}$$

where m = mass, $ds^2 = \gamma^*(g)$, the induced metrics. The image of [0,1] is the particle's virtual world-line.

String propagating in a space-time (M,g). Here $\mathfrak{P}$ = maps: $\sigma: N \to M$, where N is a surface whose image is the string's world-sheet,

$$S(\sigma) = -\frac{T}{2}\int_N \mathrm{vol}_{\sigma^*(g)} \qquad \text{(Nambu's action)} \tag{3}$$

where T is the so-called "string-tension" of dimension (length)$^{-2}$ and $\sigma^*(g)$ is the induced metric.

We shall always measure time in length units and action in Planck units (or, as physicists say, put $\hbar = c = 1$).

b. Classical equations of motion.

They are equations for stationary points of S: $\delta S = 0$. Finding their solutions or exploring their qualitative properties is the main task of classical mathematical physics.

c. Quantum expectation value and partition function.

They are given by the following formal expressions (Feynman integrals):

$$<O> = Z^{-1}\int_{\mathcal{P}} O(p)e^{iS(p)}Dp, \tag{4}$$

$$Z = \int_{\mathcal{P}} e^{iS(p)}Dp, \tag{5}$$

where $O: \mathcal{P} \to \mathbf{R}$ is an observable and Dp is a formal measure on $\mathcal{P}$.

Most problems of quantum field theory can be thought of as problems of finding a correct definition and a computation method for some or other Feynman path integral. From a mathematician's viewpoint almost every such computation is in fact a half-baked and ad hoc definition, but a readiness to work heuristically with such a priori undefined expressions as (4) and (5) is a must in this domain.

There are several standard tricks to start with. First, one tends to work with the so-called "Euclidean" versions of (4), (5), where $e^{iS(p)}$ turns into $e^{-S(p)}$. Besides "better convergence" (whatever this means), this makes explicit a basic analogy between quantum field theory in (D space, 1 time) dimensions and statistical physics in D+1 space dimensions allowing us to use the rich technical tools and insight on collective phenomena. Second, one tries to reduce (4), (5) to finite-dimensional integrals using group invariance and/or approximations. Third, one tries to reduce (4), (5) to Gaussian integrals, whose theory is the only developed chapter of infinite-dimensional integration, e.g., by means of an appropriate perturbation series.

Here is, for example, a standard heuristic explanation of the correspondence between classic and quantum laws of motion. Assume for simplicity that (with given boundary conditions) the equation $\delta S = 0$ has only one solution $p = p_0$. Assume also, by analogy with the finite-dimensional case, that the stationary phase approximation is valid, i.e., the quantum expectation value $< O > = \int_{\mathcal{P}} O(p)\, e^{iS(p)}\, Dp$ up to a universal factor and a small error coincides with $O(p_0)$. This means that quantum observables practically have their classical values on the classical path.

A necessary condition for the validity of the stationary phase approximation is that $S = S/\hbar$ be large on $\mathcal{P}$. This is in accordance with the discovery of early quantum theory that the classical regime corresponds to $\hbar \to 0$.

When a calculation of a path integral succeeds, it involves one or several limiting processes which differ from the Archimedes-Newton-Lebesgue prescription of adding up an infinity of infinitesimal contributions. In fact, such a calculation gives a finite value usually as a difference or a quotient of two (or more) infinities.

I believe that there is a message in this observation. Each level of reality we become aware of is but a flimsy foam on the surface of an infinitely deep ocean, usually called a vacuum state. It is a state of lowest energy, but its energy is infinite. We are divided by a thin film from an eternal fire, whose first tongues are the flames of the nuclear age. Will a mature string theory start a new Auto-da-Fé ?

Returning to mathematics, it sometimes happens that infinities of a concrete model apparently do not reduce to a finite number of them. A notorious example is the Einstein gravity theory (1) which for this reason is called unrenormalizable. Only after an extension to a larger picture, hopefully a stringy one, can gravity become finite.

If a theory is renormalizable (or, as happens with super-symmetric models, even a finite one), the indeterminacies in choosing "infinite constants" are resolved by referring to experimental data - values of various charges and coupling constants. Ideally, nothing like this should be allowed: a perfect theory must predict everything.

d. Operator approach.

We have described earlier the Lagrangian approach to quantization. There is an alternative approach which in various contexts is called the Hamiltonian, canonical or operator one. In those case where a sufficiently close connection can be established between them, it takes the following form.

On the space of solutions P of the classical equations of motion $\delta S = 0$ there is a natural Poisson structure, i.e., a Lie bracket on a space F of functionals on P. For example, in the classical mechanics the space of classical paths P can be identified with the phase space, since a classical motion is defined by its initial values of position and momentum. The corresponding Poisson structure is defined by the well-known symplectic form. (This basic example suggests taking for P an appropriate space of boundary values which is often done in string theory.)

A unitary representation of a subalgebra of F in a Hilbert space H defines an operator quantum description of the system. Of course, such a representation is rarely unique, and the non-uniqueness corresponds to the indeterminacy of path integrals. The so-called "geometric quantization" is a method of constructing such representations in functions on P, or sections of a bundle on P, or in an appropriate cohomology.

An expectation value $\langle O \rangle$ of an observable $O \in F$ calculated by means of a path integral should coincide with an operator expectation value of the kind $\langle \text{vac} | \rho(O) | \text{vac} \rangle$ (or $\langle \psi | \rho(O) | \phi \rangle$ for appropriate state vectors vac, ϕ, $\psi \in H$) defined by a representation ρ. In order that this make sense one should ensure that the properties of O are such that it may be considered to be a functional on $\mathfrak{P}$ uniquely defined by its restriction on $P \subset \mathfrak{P}$. An example is the locality of O meaning that O depends only on the value of $p \in \mathfrak{P}$ and some of its derivatives at a point of space-time.

In general, the path integral quantization and operator quantization should be considered as complementing each other rather then being strictly equivalent. This complementarity inherited from classical mechanics, which reappeared in the guise of the Schrödinger and Heisenberg approaches to quantum mechanics and subsequently in various guises, has dominated quantum field theory.

In the string theory, comparison of the two approaches poses most intriguing problems on the connections between modular forms on Teichmüller spaces and moduli spaces of vector bundles, on the one hand, and representation theory of the Virasoro, Kac-Moody, and similar Lie algebras, on the other hand. Before the advent of quantum field theory in this mathematical domain only genus one modular forms appeared as character series of representations, and even that seemed a mystery.

e. Symmetries.

The basic structure $(\mathfrak{P}, S)$ is often complemented by an action of a group G on $\mathfrak{P}$ leaving S invariant (or, infinitesimally, by an action of a Lie algebra $\mathfrak{A}$

which in an infinite-dimensional situation may well be non-integrable). Mathematical aspects of such a picture may have various physical interpretations of which we mention several.

Classical symmetries of a flat space-time acting upon $\mathfrak{P}$ give rise to the energy-momentum operators. Noether's theorem on conservation laws is reflected in the structure of the momentum map $\mu: P \to \mathfrak{U}^*$ where P is the space of classical motions with an invariant symplectic structure.

Local gauge symmetries in the theory of Yang-Mills fields and diffeomorphisms of the space-time of general relativity give rise to the physically indistinguishable states. Thus in this case, one should actually call $\mathfrak{P}/G$ the space of virtual paths, choose various G-invariant subspaces as spaces of quantum states, etc. String theory also provides such phenomena.

Path integral (or operator) quantization may break up classical G-invariance of the picture due to the indeterminacies in the regularization scheme.

A precise description of the resulting non-invariance is the subject matter of the theory of anomalies. In recent years it became clear that an essential aspect of anomalies reflects cohomology properties of G. Vanishing of quantum anomalies is considered as an important metatheoretical criterion of the consistency of a quantum model. It is this vanishing that led to the discovery of critical dimensions and preferred gauge groups.

Finally, a few words should be said about the conformal group. It is the group of local rescalings of a metric of space-time or string world sheet: $g_{ab} \to \to e^f g_{ab}$. The conformal invariance of a physical model implies absence of a natural scale (length, mass or energy). In the context of statistical physics this happens in the vicinity of phase transitions. Lattice approximation to path integrals, if it converges at all, does so in such a critical regime. One can conjecture that fundamental physics is governed by conformally invariant laws. Anyway, conformal invariance plays an essential role in string theory.

f. Correspondence principles.

Historically, the correspondence principle is a loosely formulated prescription for deriving classical laws of physics from quantum ones.

Modern fundamental physics is a conglomerate of theories or models, each of which is applicable at an appropriate scale or is a simplified version of a more adequate but too complex theory. All informal rules of patching together these models at the outskirts of their validity ranges may reasonably be called correspondence principles; such are various symmetry breakings and spontaneous compactifications.

If, as I believe, this openness of physics is its essential characteristic, then correspondence principles themselves may be elevated in status and can eventually become considered as physical laws acting in transition periods as *rites de passage*. For example, 26, 10 and 4 may have been successive steps of formation of our space-time from an infinite-dimensional quantum chaos during the first $10^{-?}$ seconds of the Creation.

POLYAKOV'S PARTITION FUNCTION FOR BOSONIC STRING

Let me now fill the scheme of the previous section with some concrete mathematics. I shall choose from the vast field a representative fragment, namely, a calculation of (perturbative) partition function (5) of the Polyakov bosonic string. Here is its formal description:

$$Z = \sum_{g \geq 0} Z_g,$$

$$Z_g = e^{\beta(2-2g)} \int_{\mathcal{P}_g} e^{-S_g(x,\gamma)} \, Dx \, D\gamma, \tag{6}$$

$$S_g(x, \gamma) = \int_{N_g} e\gamma^{ab} \partial_a x^m \, \partial_b x^m \mathrm{vol}_\gamma . \tag{7}$$

Here N_g is a fixed compact oriented surface of genus g, $\mathcal{P}_g = \mathrm{Map}(N_g, \mathbf{R}^d) \times$ $\times \mathrm{Met}\, N_g$, where a map $x\colon N_g \to \mathbf{R}^d$ is given by d real-valued functions on N_g and $\gamma \in \mathrm{Met}\, N_g$ is a metric $\gamma_{ab} dz^a dz^b$ written in local coordinates (z^1, z^2) on N_g, $\partial_a = \partial/\partial z_a$. Finally, β is a constant, "inverse temperature", which will not concern us here.

Informally, one should imagine the space $\mathcal{P}$ of virtual classical paths in this model as consisting of all compact parametrized Riemann surfaces in a Euclidean space-time $\mathbf{R}^d$. The Polyakov action (7) differs from the Nambu action (3) — it depends also on an intrinsic metric γ while (3) is determined by the induced metric only. However (3) and (7) lead to the same classical equations of motion $\delta S = 0$ expressing that the surface in question is a minimal one.

A different interpretation is that of quantum field theory on a two-dimensional space-time N. From this viewpoint, the action (7) describes d scalar fields x^m coupled to gravity γ.

Both in string theory and in quantum gravity, a matter of principle is a proper accounting for various topologies. Here we do it by a simple-minded summing up over all genuses g. This is precisely what gives to (6) a flavour of a perturbation series.

Now we shall fix g and try to make sense of the path integral (6). In order to do that, we start with the remark that integration over x with fixed γ is a Gaussian one. Moreover, we discover that an infinite-dimensional group $G = C \times D$ acts upon $\mathcal{P}_g$ leaving S_g invariant. Namely, it is the semi-direct product of the diffeomorphism group D of N and the conformal group C of rescalings of γ. Using this, we can try to reduce (6) to a finite-dimensional integral. Let us proceed.

Gaussian integration provides an infinite-dimensional analogue of the formula

$$\int_{\mathbf{R}^n} \exp(-1/2\,(X^n, AX^n))DX^n = (2\pi)^{n/2}(\det A)^{-1/2}, \tag{8}$$

where $DX^n = dx^1 \ldots dx^n$ and A, a positive definite symmetric operator on $\mathbf{R}^n$. With fixed γ, we put by analogy

$$\int_{\mathrm{Map}(N_g,\mathbf{R}^d)} \exp(-S_g(x,\gamma))\; Dx = (\det{}'\Delta_{0\gamma})^{-d/2}\,, \tag{9}$$

where $\Delta_{0\gamma}$ is the Laplacian of γ acting on scalar functions on N and $\det'\Delta_{0\gamma}$ means its regularized determinant defined, e.g. , by means of the zeta-function regularization: if λ_i are non-zero eigenvalues of $\Delta_{0\gamma}$, we define

$$\zeta(s) = \sum \lambda_i^{-s}, \quad \text{for } \mathrm{Re}\, s \gg 0,$$

prove that this function has a meromorphic continuation to the whole complex plane, regular at $s = 0$, and finally put

$$\det{}'\Delta_{0\gamma} = \exp(-\zeta'(0)). \tag{10}$$

Trying to untangle the limiting procedures implicit in (9) and (10) we learn several lessons. First the formal measure $\exp(-S_g(x,\gamma))Dx$ in (9) looks like a limit of

$$\mu_n = (2\pi)^{-n/2} \exp(-1/2\,(X^n, AX^n))DX^n$$

(or, rather, of $\prod_1^d \mu_n$). Therefore Dx itself has no intrinsic meaning, so that formal notation is misleading. Second, $(2\pi)^{n/2}$ furnishes (for $n \to \infty$) the simplest example of an "infinite constant" which physicists often dispose of by such catchwords as "volume of diffeomorphism group" etc. Third, the regularization of the determinant of an elliptic operator by means of (10) involves one more division by an infinite constant since the right hand side of (10) is essentially the product $\prod(\lambda_i/\lambda_i^0)$ where λ_i^0 are eigenvalues of a conveniently chosen fixed Laplacian.

This last trick is responsible for the breaking of conformal invariance of (6). In fact, if we put (9) into (6) and (forgetting β) try to define

$$Z_g = \int_{\mathrm{Met}\, N_g} (\det{}'\Delta_{0\gamma})^{-d/2} D\gamma\,, \tag{11}$$

we find out that a rescaling of γ changes the measure. In order to make this more precise we must, of course, present some explanations on the meaning of $D\gamma$. Clearly, $\mathrm{Met}\,N_g$ is a cone in the space of smooth quadratic differentials on N_g. Hence this space can be identified with the tangent space to $\mathrm{Met}\,N_g$ at any given point γ_0. We can use γ_0 and integration over N_g to define a metric on this tangent space. We then imagine that this metric defines a measure like the one denoted by Dx in (9). And although we very convincingly argued that such a measure cannot exist we can still calculate what happens with (11) after a small rescaling of γ.

Thus we find out how a conformal variation of (11) looks: it is $(26-d)$ times something.

Therefore if $d = 26$, the measure in (11) is invariant with respect to the group $C \times D$, whatever that means.

Now comes the grand finale. The point is that $C\times D \setminus \mathrm{Met}\,N_g = M_g$ is a space of finite dimension (0 for $g = 0$, 2 for $g = 1$, $6g-6$ for $g \geq 2$). It is a famous Riemann moduli space. To understand this, recall the following facts: a) to give on an oriented surface a conformal class of metrics is the same as to give a complex structure; b) there are only three connected simply connected complex Riemann surfaces — the complex plane, the complex half-plane, and the Riemann sphere; c) any complex Riemann surface is the quotient space of its universal covering space by a free action of its fundamental group. Therefore to give a point in $C \times D \setminus \mathrm{Met}\,N_g$ for $g \geq 2$ is essentially to define a representation of $\pi_1(N_g)$ in $\mathrm{PSL}(2,\mathbf{R})$. (For $g = 0, 1$ the situation is simpler.)

To understand the transition from (11) to a finite-dimensional integral, look at the following model problem. Let a connected Lie group G act on a Riemann manifold $(\tilde{M}, \tilde{h})$ leaving invariant the metric $\tilde{h}$. Let $M = G\backslash\tilde{M}$ and h the induced metric on M. If, say, everything is compact, we have $\int_{\tilde{M}} \mathrm{vol}_{\tilde{h}} = \mathrm{vol}(H) \int_M \mathrm{vol}_h$ where H is the stabilizer of a point in $\tilde{M}$ and $\mathrm{vol}(H)$ is its volume with respect to an invariant metric on G.

Using the right hand side of this formula as a definition, with $\widetilde{M}$ = Met N_g, $G = C \times D_0$, D_0 = the connected component of the identity of D, $M = T_g$ the Teichmüller covering of M_g, we can finally represent (11) as an integral over T_g(or M_g) of a concrete finite-dimensional volume form $d\pi_g$.

For example, for $g = 1$ we get the following answer:

$$Z_1 = \int_{M'_1} i/2\ d\tau \wedge d\bar{\tau}\, (\mathrm{Im}\,\tau)^{-14}\, |\Delta(\tau)|^{-2},$$

$$\Delta(\tau) = e^{2\pi i\tau} \prod_{n=1}^{\infty} (1 - e^{2\pi i\,\tau})^{24},$$

$$M'_1 = \{\tau \in \mathbf{C} \mid |\tau| \geq 1, |\,\mathrm{Re}\,\tau| \leq 1/2, \mathrm{Im}\,\tau > 0\}.$$

Recently, Belavin and Knizhnik proved that in general $d\pi_g$ equals the modulus squared of a holomorphic volume form on M_g (which is a complex orbifold). This volume form is uniquely determined up to a constant, and several more or less explicit expressions for it have been subsequently calculated.

Of course, most developments of this section mathematically were mere heuristics. Once a definite result, in our case an identification of a measure on M_g, is achieved, one can forget these heuristics altogether and work only with safely accepted tools. However, nowadays such an attitude would be counterproductive. Quantum field theory in general and specifically quantum string theory is fascinating precisely thanks to the riches of the intuitive reasoning it provides.

READING SUGGESTIONS

Entering string theory is not an easy task for the mathematician. Two quite recent publications may help him or her to get a general perspective and to choose a particular topic for deeper study: the monograph-textbook [1] and the anthology [2].

Two ICM Berkeley talks [3] and [4] were at least partly devoted to strings. Witten's talk is a beautiful initiation to quantum field theory, if ever there was one.

The Belavin and Knizhnik discovery [5] together with previous works by Quillen and Faltings led to important progress in the theory of determinant bundles [6]-[10], extending Grothendieck's Riemann-Roch theorem. A lot remains to be done in this domain which is simultaneously "the component at infinity" of arithmetical geometry [11], [12].

Hopefully, path integrals of string theory could eventually be re-interpreted in arithmetical terms via a version of Siegel-Tamagawa-Weil theory. A beautiful recent result by Shabat and Voevodsky, drawing upon previous work by Grothendieck and Bely shows that natural lattice approximations in string theory are inherently arithmetic ones (cf. [13]).

For a rich representation-theoretic part of string theory see [15], [16], [10] and many pages of [1] and [2].

REFERENCES

1. M.B. Green, J.H. Schwarz, and E. Witten, *Superstring theory*, in 2 vols. Cambridge University Press (1987).
2. J.H. Schwarz, ed., *Superstrings. The first fifteen years of super-string theory*, in 2 vols. Singapore: World Scientific, (1986).
3. E. Witten, 'Physics and geometry", Berkeley ICM talk (1986).
4. Yu.I. Manin, 'Quantum strings and algebraic curves', Berkeley ICM talk (1986).
5. A.A. Belavin and V.G. Knizhnik, 'Algebraic geometry and the geometry of quantum strings', *Phys. Lett.* **168B** (1986), 201-206.
6. D.S. Freed, 'Determinants, torsion, and strings', Preprint MIT (1986).
7. J.-M. Bismut, H. Gillet and C. Soulé, 'Analytic torsion and holomorphic determinant bundles', Preprint Orsay 87-T8 (1987).
8. P. Deligne, 'Le déterminant de la cohomologie', Preprint Princeton (1987).
9. A.A. Beilinson and Yu.I. Manin, 'The Mumford form and the Polyakov measure in string theory', *Comm. Math. Phys.* **107** (1986), 359-376.
10. A.A. Beilinson and V.V. Schechtman, 'Determinant bundles and Virasoro algebras', Submitted to *Comm. Math. Phys.* (1987).
11. G. Faltings, 'Calculus on arithmetic surfaces', *Ann. of Math.* **118** (1984), 387-424.
12. Yu.I. Manin, 'New dimensions in geometry', in: *Springer Lecture Notes in Math.* **1111** (1985), 59-101.
13. V.A. Voevodsky and G.B. Shabat, 'Equilateral triangulations of Riemann surfaces and curves over algebraic number fields', Preprint (1987).
14. D.V. Boulatov, V.A. Kazakov, I.K. Kostov and A.A. Migdal, 'Analytical and numerical study of dynamically triangulated surfaces', *Nucl. Phys.* **B275** (1986), 641-686.

15. G.B. Segal, 'Unitary representations of some infinite dimensional groups', *Comm. Math. Phys.* **80** (1986), 301-342.
16. B. Feigin and B. Fuchs, 'Representations of the Virasoro algebra', in: *Seminar on supermanifolds*, **5** (D. Leites, ed.), Rep. of Dept. of Math., Stockholm Univ., N 25 (1986).
17. P. Candelas, G. Horowitz, A. Strominger and E. Witten, 'Vacuum configurations for superstrings', *Nucl. Phys.* **B258** (1985), 46-90.

Steklov Math. Institute,
42 Vavilova, Moscow 117966,
GSP-1, USSR.

Index of names

Subject Index